D0786654

A Gift from the
Friends of the La Jolla Library

SEP 1 5 2011

Histories of Computing

HISTORIES OF COMPUTING

MICHAEL SEAN MAHONEY

Edited and with an introduction by Thomas Haigh

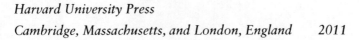

3 1336 08893 2604

Harvard University Press
Cambridge, Massachusetts, and London, England   2011

Copyright © 2011 by the President and Fellows of Harvard College
All rights reserved
Printed in the United States of America

Pages 241–242 constitute an extension to the copyright page.

Library of Congress Cataloging-in-Publication Data

Mahoney, Michael S. (Michael Sean)
    Histories of computing / by Michael Sean Mahoney ; edited and
with an introduction by Thomas Haigh.
        p.   cm.
    Includes bibliographical references and index.
    ISBN 978-0-674-05568-1 (alk. paper)
    1. Computer science—History.   2. Software engineering—
History.   3. Computers—History.   I. Haigh, Thomas, 1972–   II. Title.
    QA76.17.M34 2011
    004—dc22   2010047555

Contents

List of Figures   *vii*

Unexpected Connections, Powerful Precedents, and
Big Questions: The Work of Michael Sean Mahoney
on the History of Computing   *1*
Thomas Haigh

PART ONE   Shaping the History of Computing

1   The History of Computing in the History
of Technology   *21*

2   What Makes History?   *38*

3   Issues in the History of Computing   *42*

4   The Histories of Computing(s)   *55*

PART TWO   Constructing a History for Software

5   Software: The Self-Programming Machine   *77*

6   Extracts from *The Roots of Software Engineering*   *86*

7   Finding a History for Software Engineering   *90*

8   Boys' Toys and Women's Work: Feminism Engages
     Software   *106*

PART THREE   The Structures of Computation

9   Computing and Mathematics at Princeton in the 1950s   *121*

10   Computer Science: The Search for a Mathematical
      Theory   *128*

11   Extracts from *Computers and Mathematics:
      The Search for a Discipline of Computer Science*   *147*

12   The Structures of Computation and the Mathematical
      Structure of Nature   *158*

13   Extracts from *Software as Science—Science as Software*   *183*

      Éloge: Michael Sean Mahoney, 1939–2008   *197*
      Jed Z. Buchwald and D. Graham Burnett

      Notes   *205*
      Acknowledgments   *241*
      Index   *243*

List of Figures

Figure 1: The machine-centered view of the history of computing *58*

Figure 2: Sources of ENIAC and EDVAC *60*

Figure 3: Architecture of EDVAC *61*

Figure 4: The communities of computing *62*

Figure 5: Levels of modeling in software development *105*

Figure 6: The agenda leading from John von Neumann to the Santa Fe Institute *164*

Figure 7: The agendas of automata and formal languages *167*

Figure 8: The agendas of semantics *171*

Figure 9: A formulation of list-processing in terms of homomorphisms and a proof of the correctness of a simpler compiler *176*

Histories of Computing

Unexpected Connections, Powerful Precedents, and Big Questions

The Work of Michael Sean Mahoney on the History of Computing

Thomas Haigh

ITHIN THE HISTORY of computing community, Mike Mahoney stood out for being, as I once called him when introducing him at Colby College, an "old school" historian of science.[1] His intellectual taste was impeccable and his standards high. Few, if any, of his public lectures on computing proceeded without at least one reference to René Descartes, Isaac Newton, or Christiaan Huygens. Although he had trained himself as an expert in the early literature of theoretical computer science, he was also a skilled reader of sixteenth-century Latin and prone to lament that few students today could read Newton's *Principia Mathematica* in the original. The latter, he explained with a touch of pride, required not just a knowledge of Latin but also deep familiarity with the mathematical notation and scientific practice of the period. Yet he was by no means fusty. Instead Mahoney's vision of the history of computing reflected his openness to a broad range of approaches. Such quickness of wit and breadth of knowledge could prove intimidating to those of us whose historical knowledge stretched little beyond the late nineteenth century. His death deprived our community of a unique perspective that is unlikely to be replicated.

Mahoney's instinct was to situate the field in broader and deeper historical intellectual currents, flowing from both the history of science and the history of technology. His papers on the history of computing, gathered in this volume, reflect this intellectual virtuosity. They span an impressive range of topics but are united by a concern for the creation of new concepts, practices, and intellectual communities through the recombination and reinterpretation of old ones. In this respect his papers on historiography, software engineering, and computer science—the three areas into which I have gathered his work—ask similar questions and suggest similar answers.

Historiographic Work

Mahoney's most influential contribution to the development of the history of computing as a thriving and somewhat respectable field of scholarly labor came from his series of historiographic papers published from 1988 to 2008. This was the first area of the history of computing that Mahoney approached, but it was a topic that he returned to again and again. These papers constitute the most sustained and self-conscious examination so far attempted of the fundamental question hanging over our growing body of work: what is the history of computing a history of?

The first, "The History of Computing in the History of Technology," appeared in 1988 in *Annals of the History of Computing*. It is one of the most widely cited papers ever written on the history of computing.[2] In 1988 the "history of computing" was by no means a novel object of study: *Annals* was already in its tenth volume, and the MIT Press series History of Computing was growing rapidly. Yet work had so far been carried out almost exclusively by aging computer pioneers, joined by a growing but still tiny body of young historians whose dissertation research addressed computing. Mahoney was a tenured professor of history at one of the world's leading universities, and so his choice of the history of computing as one of his primary research interests was in itself something of a watershed moment for the field.[3]

At this time *Annals* was published by a consortium of professional computing societies, and its readership was made up largely of computer experts rather than trained historians. This was Mahoney's first published communication to this community, yet he did not waste space on false praise either of existing technical and business histories or of the willingness of historians to address vital questions related to information technology.

Instead Mahoney made it very clear that his embrace of the field reflected a belief in the inherent importance of its subject matter rather than a high opinion of anything so far accomplished by its researchers. The paper briskly asserts the centrality of computers to science, business, the "information society," and the "new concepts of information" believed to underlie it. Historians, he continued, were failing to rise to the challenge.

Despite the pervasive presence of computing in modern science and technology, not to mention modern society itself, the history of computing has yet to establish a significant presence in the history of science and technology. Meetings of the History of Science Society and the Society for the History of Technology in recent years have included very few sessions devoted specifically to history of computing, and few of the thematic sessions have included contributions from the perspective of computing.

Mahoney took just two sentences to rule out existing scholarship in the field as a source of insight.

There is a small body of professionally historical work, dealing for the most part with the origins of the computer, its invention and early development. . . . It is meant as no denigration of that work to note that it stops at the point where computing becomes a significant presence in science, technology, and society.

The force and abruptness of this dismissal is striking. Four years earlier, the young historian William Aspray had undertaken a similar survey of the history of computing in *Isis*, the most staid and venerable journal of the history of science. Aspray took a more conventional approach, dutifully surveying the existing literature and balancing his calls for attention to various pressing needs with an appreciation of existing accomplishments. Mahoney, in contrast, used vivid language to place himself and the rest of the historical profession in a position of hopeful ignorance.

Historians stand before the daunting complexity of a subject that has grown exponentially in size and variety, which looks not so much like an uncharted ocean as like a trackless jungle. We pace on the edge, pondering where to cut in.

"The question," he suggested, "is how to bring the history of computing into line with what should be its parent discipline." He concluded, "Pursued within the larger enterprise of the history of technology, the history of computing will acquire the context of place and time that gives history meaning." This was itself an interesting choice. Mahoney's training as an intellectual historian of science might have led him to dismiss the history of technology as a dull and low-status field, devoid of big ideas. He could easily have positioned the computer as a creation of science, an embodiment of mathematical logic, and the most important laboratory instrument ever invented. Indeed his work would explore these very perspectives. Yet his agenda for the history of computing was firmly rooted in the grubbier world of technology.

To sketch the outlines of a possible history of computing rooted within the history of technology, Mahoney offered questions rather than answers.

Specifically, he posed a number of "big questions" gleaned from major works in the history of technology, placing the computer in comparative perspective with earlier technologies. Echoing the main concerns of historians of technology in the 1970s and early 1980s, they were concerned particularly with the organization of work, the relationship of science and technology, the professional development of engineering knowledge, and (rather less typically for this period) the materiality of computer technology and the importance of tinkering and communities of enthusiasts to its shaping and spread.

In 1993 Mahoney made another argumentative introduction to the field in his paper "Issues in the History of Computing," presented to an audience of computer scientists gathered for a conference on the history of programming languages. Of all the Ph.D. historians involved with the history of computing, Mahoney spent the most time and effort working with technical communities to spread awareness of history and guide their historical initiatives in productive directions. As someone who has attempted to reach audiences within computer science and information science, I can testify to the difficulty involved in convincing such audiences that Ph.D. historians hold important skills, perspectives, or research questions usually lacking in the work of those with an avocational interest in the technical developments of their own fields. It is easy to cause offense and hard to change minds. Mahoney's work for such audiences is a model of its kind: charmingly self-deprecating, eminently reasonable, elegantly argued, yet unsparing in its dismissal of Whig history and antiquarianism. This stemmed, perhaps, from his genuine respect for computer scientists and deep immersion in the discipline's technical literature.

Five years on from his earlier paper, Mahoney remained unimpressed with the existing literature, most of which had been written by participants rather than trained historians. His words recall historian of science Thomas Kuhn's definition of the state of a "preparadigmatic" research field.[4] According to Mahoney, in the history of computing, sources are abundant, but "we have lots of answers but very few questions, lots of stories but no history, lots of things to do but no sense of how to do them or in what order. Simply put, we don't yet know what the history of computing is really about." Writers from the computing community were preoccupied with establishing credit for "firsts," and computing was still "invisible" within the Society for the History of Technology.

Mahoney offered three main suggestions to remedy this and drag the history of computing into a more productive relationship with the concerns and insights of more mature historical fields. Rather than focus exclusively on great men and breakthrough accomplishments, Mahoney believed that historians of computing should document and interpret common practices and expose the importance of tacit knowledge to the field. This, as he made explicit elsewhere, he took from Kuhn's arguments for the centrality to science of routine problem-solving procedures.[5] He called for general attention to the role of "commerce," in which he included both the use of computers in business and the role of research-funding agencies in shaping the evolution of computing technology. He challenged historians to use insights from historical and sociological work on the development of professions, particularly the insights of Andrew Abbott, to explore the evolution of professional identities around the computer and the demarcation of occupational jurisdictions.

Mahoney's final major contribution to the historiography of computing came with his 2005 paper "The Histories of Computing(s)." In this Mahoney offers a new kind of answer to his old question: what is the history of computing about? His answer: many different things, because it is not the history of one thing but of many. Looking back, one sees this theme even in his earliest work on the subject. In his 1988 paper, Mahoney had noted that

> The computer is not one thing, but many different things, and the same holds true of computing. There is about both terms a deceptive singularity to which we fall victim when, as is now common, we prematurely unite its multiple historical sources into a single stream, treating Charles Babbage's analytical engine and George Boole's algebra of thought as if they were conceptually related by something other than twentieth-century hindsight.

We have tended, he observed in 2005, to take a "machine-centered" view of its history, "tracing its origins back to the abacus and the first mechanical calculators and then following its evolution through the generations of mainframe, mini, and micro." Scholars in the history of technology have been increasingly turning their attention to technology in use, and given the flexibility of computer technology it seems particularly appropriate here. Yet historical periodization around machine generations and depiction of a family tree of great machines going back to ENIAC are built around hardware technologies and computer producers. If one is interested in users and applications this makes no sense at all.

In this paper Mahoney applied the same analytical tools to the conceptualization of the history of computing that he had been applying to understand the construction of computer science and software engineering. In particular he was interested in the hunt for precedents and the ways in which historical analogies can become locked into the structure of a field. His illustrations for this paper provide an impressively direct statement of his concept. On the one hand we see the hourglass shape of the "history of computing," in which earlier technologies shaped ENIAC and its compatriots, which in turn give birth to subsequent generations of computers with an ever-wider range of applications (see fig. 1 in chapter 4). On the other lie the "histories of computing(s)," in which social practices in a variety of communities (business, scientific fields, industrial production, and so forth) are gradually restructured around the possibilities of computer technology (see fig. 4 in chapter 4).

Mahoney did a lot to legitimate "the history of computing" as the identity of a respectable area of study, so it is striking that in his final historiographic work he turned away from that particular frame in favor of an explicitly plural alternative centered on users and practices. In this paper Mahoney identifies and acknowledges the influence of a few streams of promising work in these areas, reflecting the growing maturity of historical scholarship on computing since the publication of his 1988 paper. This is almost unique in his published work. Mahoney's papers rarely cite work by others in the history of computing and give the impression of a lone historian fighting a brave battle against an overwhelming mass of primary sources and urgent research questions. That impression is misleading. In fact he was deeply engaged with other scholars within our field, a keen participant on panels and in workshops, and consistently interested in the work of graduate students.

Mahoney's argument in this paper tacks gracefully to address two potentially contradictory objectives. The first is his urge to strip away the rhetoric of revolution in which the computer has been enveloped since its invention. Computers, we have been told for more than sixty years now, are so much more powerful and important than any earlier technologies that it would be at best pointless, and at worst dangerous, to use history as a guide. Mahoney rightly tied this rhetoric to the discredited concept of technological determinism and the idea that technologies "impact" upon a society. Instead his urge was always to compare computers with other technologies, from windmills to cars.

One would be surprised if a serious historian of technology, particularly one with Mahoney's broad range of interests, took any other position. Discovering unexpected parallels between phenomena ancient and modern is, after all, our intellectual party trick. In contrast, if a historian accepts uncritically the assumption of his or her historical subjects that their group, nation, or technology has a unique character and requires special kinds of explanation, then the historian's work is liable to be dismissed as simpleminded. Among academic historians, for example, to suggest that a colleague has implicitly accepted the idea of American exceptionalism is not a compliment. Analysts of computing are likewise vulnerable to naive endorsement of what I like to call computer exceptionalism, rather than seeing the computer as one important technology among many.

Yet Mahoney's second impulse ran in the opposing direction. Like many of us, he suspected that deep down, there really was something special about computer technology. Why, after all, had he been drawn to spend decades immersing himself in its mysteries? So immediately after proving his historical credentials by denouncing the rhetoric of the computer revolution, Mahoney pivoted to explore just what makes the computer, after all, an exceptional technology.

> [The stored program digital computer] could . . . do things no earlier machine had been able to do, namely, make logical decisions based on the calculations it was carrying out and modify its own instructions. . . . Capable of calculating any logical function, it could become anything but was in itself nothing (well, as designed, it could always do arithmetic).

Invocation of the computer's special powers as a universal machine is a widely used rhetorical flourish, but in this case Mahoney challenges us to combine it with his earlier dismissal of the idea of a computer revolution. Universality should be the starting point of a serious comparative examination of the computer as a technology, not a devastating first strike on the idea that computing has important historical precedents. This explains Mahoney's particular interest in software as the thing that makes a computer a computer. Computer hardware is just another kind of complex technology. Computer software is something unique, a self-executing text bridging Mahoney's interests in mathematics and machinery. The immateriality of software, however, posed new challenges to the historian.

Software Engineering

Mahoney explored these challenges in his work on the history of software and software engineering. This was his first major area of interest in the history of computing, and at one point he planned a book on the topic.[6] An overview of his concern with software is given in the short article "Software: The Self-Programming Machine." This explains his particular interest in systems software as the crucial innovation making it practical to program the computer for a wide range of applications. As he put it, "if application software is about getting the computer to do something useful in the world, systems software is about getting the computer to do the applications programming."

Mahoney called many times for a detailed examination of the actual practices of computing, and particularly those of programming, but he never took up his own invitation. This remains an underdeveloped topic within the history of computing, in contrast to work in other areas of science studies and the history of technology.[7] The papers he wrote on the history of software and software engineering are closer in style, content, and concerns to his historiographic work than they are to his detailed descriptions of theoretical computer science. The development of software engineering is sketched or alluded to rather than being presented in depth as a narrative. Though a number of prominent computer scientists and system software developers are quoted at length, we never really learn who these people are, how their ideas were formed and around what kind of practice, what their broader agendas were, or whether anything they said had a direct influence on the subsequent development of programming work. Neither do we get a sense of the institutional and disciplinary interests involved. The development of specific programming methodologies in the 1970s and 1980s is neglected here, as it generally has been in the work of others concerned with the same issues.

Instead Mahoney launched a sustained examination of precedent and the construction of historical understandings within technical communities. This theme is central to his two historiographic papers of the 1990s, his work on the history of software engineering, and much of his work on the emergence of theoretical computer science. Comparison of the computer with earlier technologies was not only, he suggested, a way of finding some meaningful research questions for historians grappling with its mysteries. It was also the means by which historical actors themselves had understood the potential of the new technology and shaped their own plans for its development. The theme received its most direct statement in his 1990 paper "The Roots of Software Engineering": "Nothing is really unprecedented. Faced with a new situation, people liken it to familiar ones and shape their responses on the basis of the perceived similarities."

"Finding a History for Software Engineering," the most mature statement of his work in the area, embeds this insight into the very structure of the paper. He identifies three intersecting agendas for the development of software engineering: applied science, mechanical engineering, and industrial engineering.

The Ford Model T appears repeatedly as an object of comparison with the computer in Mahoney's work. He observed that his historical actors were themselves invoking the Model T as an inspiration for the mass production of software, producing programs on an assembly line rather than by traditional craft practices. Mahoney discussed this both in "Issues in the History of Computing" (where he looked at computer pioneer Grace Hopper's use of the metaphor) and in his papers on the origins of software engineering. In the former he wrote that "there is history built into it, as there is in the notion of engineering itself. . . . There were other ways to think about writing programs for computers." The choice of a precedent determines the assumptions behind a field and so shapes its further direction. Ignorance of this history thus deprived practitioners of an understanding of the assumptions built into their own work practices. Mahoney himself was aware of the dangers of misinterpreting such historical icons, particularly as the computer had not yet accomplished any similar transformation of daily life and work in America. His insights here were shaped by his teaching more than by his research, particularly his commitment to undergraduate seminars exploring crucial technologies in their social contexts and his instructional concept of "reading a machine" for which he took the Model T as his paradigm.[8]

The other main topic in these papers is the "software crisis," an idea Mahoney first referred to in "The History of Computing in the History of Technology" and made central to his work on software engineering. This has been the most widely discussed topic in the history of software over the past twenty years.[9] Mahoney tied the attractiveness of software engineering as a new identity to the prevalence of a sense of crisis around large-scale programming projects in the late 1960s. He drew attention to the 1968 NATO Conference on Software Engineering as a pivotal moment in this process, both in demarcating the nature of that crisis and in building an elite consensus that new approaches were needed. Drawing on his interest in integrating the history of computing with broad themes in the history of technology, he invoked the specter of Frederick W. Taylor and his laws of scientific management as the animating force behind much of this discussion of standardizing methods for the production of software. While other scholars, most notably Donald McKenzie, worked on the topic around the same time, it is clear that Mahoney's conception of the crisis has been extremely influential on the work of subsequent scholars.[10]

Mahoney had a deep interest in the history of one particular software system: UNIX. As he began to immerse himself in the world of computing science during the 1980s, many key members of the team responsible for UNIX were still working in the flagship Bell Labs facility in Murray Hill, not far from Princeton. UNIX provided an exceptionally clear example of the link connecting the social context in which a software system was developed, its technical architecture, and its strengths and weaknesses. Mahoney conducted several interviews during the late 1980s with the creators of UNIX and later enlisted Princeton students to work further on this "Unix Oral Histories Project," material from which is available on-line, though it was never fully edited or formally published. When Linux surged in popularity during the late 1990s, Mahoney adopted it for his personal use and was often seen wearing a promotional red hat from the Linux vendor of the same name.

His only published discussion of UNIX comes in the paper "Boys' Toys and Women's Work," most of which is included in this volume. It begins in a similar vein to his historiographic work, dutifully identifying questions from the social history of technology relevant to the history of women in computing. For a moment he seemed in danger of falling into the very trap he later identified in "The Histories of Computing(s)" by assuming a single history of computing beginning with ENIAC and a single professional "field" from which the disappearance of women over the course of the 1950s must be explained. But he then made an abrupt turn into detailed critique of two papers on the gendered sociology of programming and in doing so he made his own arguments about the assumptions and social organization embedded in UNIX. According to Mahoney, the toolbox philosophy guiding UNIX was deeply grounded within computer science, supported what would later be called rapid prototyping, and was deliberately based on a "hacker" philosophy.

The last of Mahoney's papers on software is "What Makes the History of Computing Hard and Why Does It Matter?" published in *IEEE Annals of the History of Computing*. A revised version of "The Histories of Computing(s)," it was delivered at a 2007 Austrian workshop on "Methodic and Didactic Challenges of the History of Informatics," a rather ponderous title apparently chosen for its ability to be shortened to "Medichi." Again Mahoney had given himself the challenge of explaining good historical practice to an audience of distinguished computer scientists with an avocational interest in the history of their field. His choice of material underlines the historiographic character of his work on software history. For this venue he removed the introductory reflections on the role of the historian and the concluding section on the importance of computing to the humanities, stripped out some of the historiographic detail, made minor changes to the text, and added a new concluding section called "Legacy Software as History." Here Mahoney made the argument that history is important to computer scientists and information systems workers because they are working with programming and languages and operational systems that, thanks to the entrenched power of legacy systems and dominant standards, are shaped by choices made decades ago. His final paper concludes, "Even as computer scientists wrestle with a solution to the problem, they must live with it. It is part of their history, and the better they understand it, the better they will be able to move on from it."[11]

Theoretical Computer Science

Mahoney's main project of the 1990s was a book called *The Structures of Computation: Mathematics and Theoretical Computer Science, 1950–1970*. It was to have explored the emergence of theoretical computer science as a discipline, which he believed to have taken place around efforts to develop a mathematical model of the digital computer. Introducing Mahoney at the Second ACM History of Programming Languages Conference in 1993, programming language expert Jean Sammet stated that this book was "nearing completion" and that he would then return to his project on software engineering.[12] But in reality progress was slow, and by the early 2000s Mahoney was no longer confident that the book would ever be finished. Though his Web site continued to list it as a work in progress, he never produced a complete draft.

This was the first, and so far the only, attempt to write a reasonably broad intellectual history of computer science. Several historians have explored the institutional development of funding agencies and their role in shaping work in the field.[13] As for the science itself, historians and journalists have produced little more than a small shelf of books on Alan Turing, now a minor gay icon, and a smaller outcrop of work on John von Neumann.[14] Computer scientists have written a range of obituaries, appreciations, memoirs, historical literature surveys, histories of leading academic programs, and narratives of work in their specialist subfields, but the coverage provided by these is very patchy and none make a serious attempt at the history of the discipline as a whole.[15]

Mahoney's published papers on the topic broke new ground, providing detailed and immersive glimpses of the history of theoretical computer science. As with his work on the history of mathematics and natural science, his dominant concern was to reconstruct the work of leading scientists within the intellectual context in which it took place. As a result these papers are perhaps the most compressed and demanding to be found in the entire history of computing literature. His papers on historiography and software were crafted with specific audiences in mind and led these audiences expertly toward particular conclusions, easing the journey with wit and a conversational style. In contrast, his work on theoretical computer science plunges readers into a technical world without very much in the way of context or signposting. These papers do not explicitly frame their argument within contexts of questions from the history of science literature, or summarize for the reader any broader contributions to our understanding of the evolution of science. Indeed Mahoney often seemed to prefer his audiences to derive their own sense of precisely what his argument was, unusual in an age where we expect a thesis to be stated clearly at the beginning of each paper and then again at its end.

Between 1992 and 2008 Mahoney published seven papers in this area. None is wholly distinct from the others in argument, subject matter, or evidence, yet each includes unique insights. The papers focus on the 1950s and 1960s, with flashbacks to earlier developments in mathematics and jumps forward to more recent developments in computer modeling and simulation. The same names recur: Christopher Strachey, John McCarthy, Alonzo Church, Marvin Minsky, and John von Neumann. As his short paper "Computing and Mathematics at Princeton in the 1950s" makes clear, for him this was, like UNIX, a local story (its conclusion was "None of this happened in Princeton in the 1950s, but it began there.") Read together, they constitute a set of variations on a theme, with shared passages and familiar refrains combined in novel ways as he experimented with one or another framing argument or historical angle. A core set of ideas and events appear again and again. In particular, we see the way Mahoney relied on the exposition of lengthy quotations to structure his work. The quotations reappear even as the exposition around them shifts. Mahoney uses the same pair of observations from John McCarthy on astronomy during the scientific revolution as a model for the use of mathematics in computer science in four of his major papers on computer science. Tony Hoare's charming admission that the mathematical nature of programs must be asserted as "self-evident" because "nothing is really as I have described it, neither computers nor programs nor programming languages nor even programmers" appears twice, as do passages from Scott, Strachey, and several others. A passage of von Neumann, concluding with the observation that the theory of automata must be "from the mathematical point of view, combinatory rather than analytical," appears in at least six of Mahoney's published papers and chapters. (Only one quoted passage is comparably ubiquitous in his historiographic work: all three of Mahoney's major papers in this area include Henry Ford's description of the engineer as tinkerer).

For this volume I have aimed to strike a balance between avoiding duplication and illuminating the development of Mahoney's thought by publishing three of these papers in their entirety and portions of two more. Given their daunting density, the reader will benefit from having more than one treatment of this story. My experience was that Mahoney's talk on the topic was much less intimidating by the third time

I heard it. Whatever difficulties he encountered in producing the book he had planned, he could surely have published a useful volume merely by integrating the material in these papers, providing more of the context, and incorporating in the narrative some of the background knowledge he assumes of the reader.

"Computer Science: The Search for a Mathematical Theory" is the most complete statement of Mahoney's work on the topic, outlining the formerly disparate areas of knowledge (the theory of comparable functions, theory of automata, coding theory, formal language theory, and lambda calculus) that were combined to create the new discipline as well as briefly sketching the process of synthesis by which this occurred. As with "Finding a History for Software Engineering," Mahoney is acutely conscious of the process by which precedents were searched for and claimed. His earlier paper "Computers and Mathematics: The Search for a Discipline of Computer Science" focuses much more on the 1960s, particularly on vital contributions of British computer scientist Christopher Strachey to the adoption of lambda calculus as "a metalanguage for specifying and analyzing the semantics of programming language." In these papers his primary concern is on the connections made between followers of different mathematical traditions and the breadth of influence on the eventual shape of the field. Well-known figures such as Turing and linguist Noam Chomsky play their part in this narrative, but so do others such as mathematician Marcel P. Schützenberger, whose contributions are much less widely appreciated within computer science.

The use of charts in these papers to demonstrate connections and flows of influence among people, concepts, theories, and systems was unusual in historical practice. They function effectively to diagram the complex structure of Mahoney's understanding of the field's compilation, hinting at parts of his broader intended narrative he never fully developed. Seeing "Shannon," "Turing," and "Chomsky" on the same slide with dozens of arrows and intervening nodes makes a powerful impression, emphasizing his interest in connections rather than detailed studies of individual topics (see fig. 7 in chapter 12). The charts have a very similar style to the "traces" produced in the ACM SIGSOFT Impact Project, to which Mahoney was an (unpaid) historical consultant.[16] He convinced project members that historical influence was a complex process in which seminal work had its impact indirectly and in combination with other streams.[17] In turn he benefited from refinements and formalization of this graphical technique. This is an excellent example of the two-way flows of knowledge and practice he achieved in his dealings with the technical communities of computing.

Mahoney's later papers drove this stream of research in an unexpected direction, perhaps marking the divergence of his personal understanding of the historical significance of theoretical computer science from the intended structure of his fading book project. Instead of zooming in to flesh out the details in his story of the genesis of computer science, he pulled back for a panoramic view in which the 1950s and 1960s were merely a part of a broader narrative stretching back to the scientific revolution and forward into the future of scientific practice. Mahoney's work had long represented the central challenge of theoretical computer science as the production of mathematic models of the behavior of computer programs. In "The Structures of Computation and the Mathematical Structure of Nature" he recapitulated this story, with particular attention to cellular automata and lambda calculus. Again Mahoney returned to his characteristic themes of the quest for precedent and the post hoc assemblage of formerly disparate traditions as the basis for a new discipline. This time, however, the story is framed within the broader history of applied mathematics and applied science. Mahoney explored connections between the use of mathematical tools to model computer programs and the use of computers to model nature, something explicit in early work in the field (particularly that of von Neumann and Turing) and now increasingly visible thanks to the adoption of computer models in many scientific fields and an enthusiasm for building experimental worlds inside the machine.

This is an interesting perspective, as traditional mathematical modeling, particularly the approximate solution of large systems of equations, has drifted to the margins of computer science. Numerical analysis was a major part of what became computer science in the 1950s and early 1960s but was largely pushed to the margins of curricula and professional societies by the 1970s in favor of more theoretical and disciplinarily specific topics (such as those Mahoney himself wrote about). But here Mahoney argued convincingly for ties between new kinds of simulation and modeling and computer science.

Mahoney was fascinated by the artificial life movement of the 1990s and with the application of lambda calculus to biological models. He twice taught a graduate seminar with Angela Creager on "Computers and Organisms." This paper gave him a way of integrating his work on mathematical practice during the scientific revolution with his interest in theoretical computer science. Classical science had relied in large part on geometry as a tool for modeling the natural world. The seventeenth century saw a major upheaval as scientists such as Newton and Huygens shifted toward what Mahoney called "an algebraic mode of thought" in which "new means," primarily those of calculus, "had changed not only the techniques of solution but also the very manner of posing problems."[18]

Algebra allowed people to reason mathematically about the internal behavior of the system being modeled rather than relying on geometric analogy. Mahoney came to believe that the widespread adoption of computer simulation was reversing this historic shift. He had devoted much of his career to exploring the initial reshaping of applied science around algebra and now saw an opportunity to explore its eventual replacement. We have, he argued, no suitable mathematical tools with which to reason about the internal functioning of a computer simulation or verify its correspondence with the natural system being modeled. So we are returning to a world in which models can be evaluated only experimentally according to their fit with observed reality. The quest of theoretical computer science to provide a tractable mathematical description of the behavior of any given program thus has much broader implications for the evolution of applied science in the twenty-first century. Mahoney's clearest statement of the idea came in a paper not included in this volume, "Calculation—Thinking—Computational Thinking: Seventeenth-Century Perspectives on Computational Science." It concluded:

> Today we confront the question of whether the computer, the newest and leading medium of scientific thought, can be comprehended mathematically, i.e., in some way algebraically or analytically. If so, then it will be viewed as the newest chapter of a history that began in the seventeenth century with the beginning of algebraic thought. If not, then perhaps fifty years from now someone will be giving a lecture on the topic of "The End of Algebraic Thought in the Twentieth Century." [19]

His paper "Software as Science—Science as Software" provides an earlier treatment of the same materials as "The Structures of Computation," including the relationship of computer science to scientific simulation. The paper is therefore presented here as a set of extracts rather than in its entirety. The particular framing of this version provides new insights. Mahoney gives his most detailed elaboration of the concept of a disciplinary research agenda, defined in several of his papers as what "its practitioners agree ought to be done, a consensus concerning the problems of the field, their order of importance or priority, the means of solving them, and perhaps most importantly, what constitutes a solution. Becoming a recognized practitioner means learning the agenda and then helping to carry it out." As he further explores the concept of an agenda, Mahoney comes closer here than in his other versions of the story to joining the intellectual and institutional developments of computer science. Because of the context in which the work was presented, he also included some new material on software as a concept and the relationship of software engineering to computer science.

For many of his readers, particularly those in technical fields, the concept of a disciplinary agenda provided a new way of thinking about their history. This perspective is clearly shaped by the work of Kuhn, who included these topics as part of the "paradigm" (later clarified as part of the "disciplinary matrix" erected around the paradigm itself).[20] When Mahoney first presented "Software as Science—Science as Software," I was charged with summarizing the discussion that followed. Making a comparison between their approaches, I learned from Mahoney's reply that Kuhn had been his primary dissertation adviser. This was a source of pride for Mahoney. In *The Structure of Scientific Revolutions,* one of the most broadly influential books of its era, Kuhn argued that no direct objective comparison between major theories was possible because each constructed a different version of the world and gives rise to a different set of concerns and practices. His historical work on the Copernican revolution and early development of quantum mechanics stressed the need to understand historical actors in the context of their own traditions and understandings of the world, rather than evaluating their work according to its fidelity to our current methods. Kuhn was a seminal advocate of the idea of science as a social process, in which rival scientific traditions vied for supremacy and institutional advantages such as the control of textbooks or leading research institutions were crucial in setting the direction of a field. But his concern was purely with social dynamics among scientists themselves, rather than with the kinds of broader social influences on scientific practice explored in his wake by others (including, but not limited to, class, gender, military needs, government policy, industrial developments, national styles, and race).

Rereading these papers on theoretical computer science, I was struck by the extent of Kuhn's influence on Mahoney's work. They are about people as well as ideas, and Mahoney was interested in their intellectual backgrounds, their institutional ties, their patterns of collaboration, and their personal styles of science. Like Kuhn, he saw the scientific community, rather than the individual idea or isolated researcher, as the natural unit of study. Also like Kuhn he was attentive to social practices within scientific elites but not to the relations of science to broader society. This is a striking contrast with Mahoney's proposed agendas for the history of computing, though of course one cannot take every worthy perspective within a single paper or even a single career.

Kuhn's great theme was cognitive revolution in science. He insisted that the history found in scientific textbooks was distorted by presenting the work of earlier scientists within present-day conceptual frameworks and disciplinary traditions. The task of the historian was to challenge this and reconstruct the alien subjective world in which their work actually took place. Mahoney approached the history of computer science in a similar way, influenced also by his own work on the scientific revolution. He was concerned above all else with the transmission and evolution of research agendas from one generation of scholars to the next. He wrote as if he had in mind an audience for whom the broad outline of early computer science, its theorists, ideas, and milestones, were already familiar. Much is alluded to rather than explained. Instead Mahoney worked to challenge our presumed assumptions about how all these things fit together, disrupting and rearranging a received narrative that has, alas, never been written in the first place (though it was surely part of the folklore of the next generation of computing researchers). This is, of course, very much the way in which people write about early modern science, in which men such as Newton and Galileo have been studied for centuries and yet continue to provide material for generation after generation of historians. It also reflects a style of teaching found in graduate history of science seminars at Princeton, in which students are assigned substantial mounds of historical scientific papers and other primary sources for each week of discussion and expected to grapple for themselves with their complexities (or learn to skim, which Mahoney called "reading aggressively," and bluff convincingly).

His papers on the history of computer science are unmistakably challenging. If, as I have suggested, he wrote them for an audience that does not yet exist, then we may be unable to fully evaluate them for several decades until a conventional history of computer science, whose assumptions Mahoney sought to challenge, has finally been assembled. I suspect that they will retain their vitality well. In the interim, my advice is to read them carefully and repeatedly until the pieces begin to fall into place. More than any of Mahoney's other work, they demonstrate the remarkable effort he made to get inside the scientific world he was writing about and the irreplaceable role he filled within the history-of-computing community as our connection to the proudest traditions of the history of science.

Summary

Mahoney's work on the history of computing stands apart from that of his peers in many ways. His papers are skillfully constructed and patiently polished. Mahoney was not just the most erudite author in our field but, sentence for sentence and paragraph for paragraph, the liveliest. His body of work was ultimately smaller than that of other important historians of computing, but he crammed the most ideas into each paragraph. He delved more deeply into the intellectual content of computer science, and he took seriously the historical efforts of its practitioners while never suppressing his own historical instincts. He approached the subject with the instincts of a traditional historian of science, working with a time span of centuries, whereas most of us think in decades.

This work exemplifies three virtues above all else. The first of these is his focus on connections: between disciplines, across historical time periods, between theory and practice, and between traditions and schools of thought. Mahoney never published a sustained investigation of any particular event, person, institution, or technology related to the history of computing, but he was acutely aware of the structures that held them together. These insights informed his historiographic work, particularly his embrace of the idea of the "histories of computing(s)." The second is his concern for precedents, not just those imposed by historians but also, and more important, the selection of precedents by our historical actors and their selective use to shape the subsequent development of science and technology. The third is his commitment to identifying the big questions others might overlook, a skill we see applied in his historiographic papers laying out the issues in the history of computing as well as in his own research on topics such as the relationship of computer simulation to scientific knowledge. The questions he raised will long outlive the answers he gave, and both are outliving him. This, surely, is the ultimate mark of success for a mature scholar entering a young field.

PART ONE

Shaping the History of Computing

The History of Computing in the History of Technology

INCE WORLD WAR II "information" has emerged as a fundamental scientific and technological concept applied to phenomena ranging from black holes to DNA, from the organization of cells to the processes of human thought, and from the management of corporations to the allocation of global resources. In addition to reshaping established disciplines, it has stimulated the formation of a panoply of new subjects and areas of inquiry concerned with its structure and its role in nature and society. Theories based on the concept of "information" have so permeated modern culture that it now is widely taken to characterize our times. We live in an "information society," an "age of information." Indeed, we look to models of information processing to explain our own patterns of thought.[1]

The computer has played the central role in that transformation, both accommodating and encouraging ever-broader views of "information" and of how it can be transformed and communicated over time and space. Since the 1950s the computer has replaced traditional methods of accounting and record keeping by a new industry of data processing. As a primary vehicle of communication over both space and time, it has come to form the core of modern information technology. What the English-speaking world refers to as "computer science" is known to the rest of Western Europe as *informatique* (or *Informatik* or *informatica*). Much of the concern over information as a commodity and as a natural resource derives from the computer and from computer-based communications technology. Hence, the history of the computer and of computing is central to that of information science and technology, providing a thread by which to maintain bearing while exploring the ever-growing maze of disciplines and subdisciplines that claim information as their subject.[2]

Despite the pervasive presence of computing in modern science and technology, not to mention modern society itself, the history of computing has yet to establish a significant presence in the history of science and technology. Meetings of the History of Science Society and the Society for the History of Technology in recent years have included very few sessions devoted specifically to the history of computing, and few of the thematic sessions have included contributions from the perspective of computing. There is clearly a balance to be redressed here.

This status of the history of computing within the history of technology surely reflects on both parties, but the bulk of the task of redress lies with the former. A look at the literature shows that, by and large, historians of computing are addressing few of the questions that historians of technology are now asking. It is worthwhile to look at what those questions are and what form they might take when addressed to computing. The question is how to bring the history of computing into line with what should be its parent discipline. Doing so will follow a two-way street: the history of computing should use models from the history of technology at the same time that we use the history of computing to test those models. In some aspects, at least, computing poses some of the major questions of the history of technology in special ways. Each field has much to learn from the other.

Computing's Present History

Where the current literature in the history of computing is self-consciously historical, it focuses in large part on hardware and on the prehistory and early development of the computer. Where it touches on later developments or provides a wider view, it is only incidentally historical. A major portion of the literature stems from the people involved, either through regular surveys of the state and development of various fields and compilations of seminal papers, or through reminiscences and retrospectives, either written directly or transcribed from their contributions to conferences and symposia. Biographies of men or machines—some heroic, some polemical, some both—are a prominent genre, and one reads a lot about "pioneers." A few corporate histories have appeared, most notably *IBM's Early Computers*, but they too are in-house productions.[3]

This literature represents for the most part "insider" history, full of facts and firsts. While it is firsthand and expert, it is also guided by the current state of knowledge and bound by the professional culture. That is, its authors take as givens (often technical givens) what a more critical, outside viewer might see as choices. Reading their accounts makes it difficult to see the alternatives, as the authors themselves lose touch with a time when they did not know what they now know. In the long run, most of this literature will become primary sources, if not of the development of computing per se, then of its emerging culture.

From the outset, the computer attracted the attention of journalists, who by the late '50s were beginning to recount its history. The result is a sizable inventory of accounts having the virtues and vices of the journalist's craft. They are vivid, they capture the spirit of the people and of the institutions they portray, and they have an eye for the telling anecdote. But their immediacy comes at the price of perspective. Written by people more or less knowledgeable about the subject and about the history of technology, these accounts tend to focus on the unusual and the spectacular, be it people or lines of research, and they often cede to the self-evaluation of their subjects. Thus the microcomputer and artificial intelligence have had the lion's share of attention, as their advocates have roared a succession of millennia.

The journalistic accounts veer into another major portion of the literature on computing, namely what may be called "social impact statements." Often difficult to distinguish from futurist musing on the computer, the discussions of the effects of the computer on society and its various activities tend on the whole to view computing apart from the history of technology rather than from its perspective. History here serves the purpose of social analysis, criticism, and commentary. Hence much of it comes from popular accounts taken uncritically and episodically to support non-historical, often polemical, theses. Some of this literature rests on a frankly political agenda; whether its models and modes of analysis provide insight depends on whether one agrees with that agenda.

Finally, there is a small body of professionally historical work, dealing for the most part with the origins of the computer, its invention and early development. It is meant as no denigration of that work to note that it stops at the point where computing becomes a significant presence in science, technology, and society. There historians stand before the daunting complexity of a subject that has grown exponentially in size and variety, which looks not so much like an uncharted ocean as like a trackless jungle. We pace on the edge, pondering where to cut in.[4]

The Questions of the History of Technology

The state of the literature in the history of computing emerges perhaps more clearly by comparison (and by contrast) with what is currently appearing in the history of technology in general and with the questions that have occupied historians of technology over the past decade or so. Those questions derive from a cluster of seminal articles by George S. Daniels, Edwin T. Layton, Jr., Eugene S. Ferguson, Nathan Rosenberg, and Thomas P. Hughes, among others. How has the relationship between science and technology changed and developed over time and place? How has engineering evolved, both as an intellectual activity and as a social role? Is technology the creator of demand or a response to it? Put another way, does technology follow a society's momentum or redirect it by external impulse? How far does economics go in explaining technological innovation and development? How do new technologies establish themselves in society, and how does society adapt to them? To what extent and in what ways do societies engender new technologies? What are the patterns by which technology is transferred from one culture to another? What role do governments play in fostering and directing technological innovation and development? These are some of the "big questions," as George Daniels once put it. They can be broken down into smaller, more manageable questions, but ultimately they are the questions for which historians of technology bear special responsibility within the historical community. They are all of them questions which can shed light on the development of computing while it in turn elucidates them.[5]

A few examples from recent literature must suffice to suggest the approaches historians of technology are taking to those questions. Each suggests by implication what might be done in the history of computing. A spate of studies on industrial research laboratories has explored the sources, purposes, and strategies of organized innovation, invention, and patenting in the late nineteenth and early twentieth centuries, bringing out the dynamics of technological improvement that Rosenberg suggested was a major source of growth in productivity. In *Networks of Power* Thomas P. Hughes has provided a model for pursuing another suggestion by Rosenberg, namely the need to treat technologies as interactive constituents of systems. Developments in one subsystem may be responses to demands in others and hence have their real payoffs there. Or a breakthrough in one component of the system may unexpectedly create new opportunities in the others, or even force a reorganization of the system itself.[6]

In detailed examinations of one of the "really big questions" of the history of American technology, Merritt Roe Smith and David A. Hounshell have traced the origins of the "American System" and its evolution into mass production and the assembly line. Both have entered the workshops and factories to reveal the quite uneven reception and progress of that system, never so monolithic or pervasive as it seemed then or has seemed since. Daniel Nelson and Stephen Meyer have entered the factory floor by another door to study the effects of mass production on the workers it organized.[7]

Looking at technology in other contexts, Walter McDougall has anatomized the means and motivation of government support of research and development since World War II, revealing structures and patterns that extend well beyond the space program. Behind his study stands the ongoing history of NASA and of its individual projects. From another perspective, David F. Noble has examined the "command technology" that lay behind the development of numerically controlled tools. At a more mundane level, Ruth Cowan has shown how "progress is our most important product" often translated into *More Work for Mother*, while her own experiments in early nineteenth-century domestic technology have brought out the intimate relationship between household work and family relations.[8]

In the late 1970s Anthony F. C. Wallace and Eugene Ferguson recalled our attention to the non-verbal modes of thought that seem more characteristic of the inventor and engineer than does the language-based thinking of the scientist. Brooke Hindle's study of Morse's telegraph and Reese Jenkins's recent work on the iconic patterns of Edison's thought provide examples of the insights historians can derive from artifacts read as the concrete expressions of visual and tactile cognition, recognizing that, as Henry Ford once put it,

> There is an immense amount to be learned simply by tinkering with things. It is not possible to learn from books how everything is made—and a real mechanic ought to know how nearly everything is made. Machines are to a mechanic what books are to a writer. He gets ideas from them, and if he has any brains he will apply those ideas.[9]

The renewed emphasis on the visual has reinforced the natural ties between the historian of technology and the museum, at the same time that it has forged links between the history of technology and the study of material culture.

The Tripartite Nature of Computing

Before trying to translate some of the above questions and models into forms specific to the history of computing, it may help to reflect a bit on the complexity of the object of our study. The computer is not one thing, but many different things, and the same holds true of computing. There is about both terms a deceptive singularity to which we fall victim when, as is now common, we prematurely unite its multiple historical sources into a single stream, treating Charles Babbage's analytical engine and George Boole's algebra of thought as if they were conceptually related by something other than twentieth-century hindsight. Whatever John von Neumann's precise role in designing the "von Neumann architecture" that defines the computer for the period with which historians are properly concerned, it is really only in von Neumann's collaboration with the ENIAC team that two quite separate historical strands came together: the effort to achieve high-speed, high-precision, automatic calculation and the effort to design a logic machine capable of significant reasoning.[10]

The dual nature of the computer is reflected in its dual origins: hardware in the sequence of devices that stretches from the Pascaline to the ENIAC, software in the series of investigations that reaches from Leibniz's combinatorics to Turing's abstract machines. Until the two strands come together in the computer, they belong to different histories, the electronic calculator to the history of technology, the logic machine to the history of mathematics, and they can be unfolded separately without significant loss of fullness or texture. Though they come together in the computer, they do not unite. The computer remains an amalgam of technological device and mathematical concept, which retain separate identities despite their influence on one another.[11]

Thus the computer in itself embodies one of the central problems of the history of technology, namely, the relation of science and technology.[12] Computing as an enterprise deepens the problem. For not only are finite automata or denotational semantics independent of integrated circuits; they are also linked in only the most tenuous and uncertain way to programs and programming, that is, to software and its production. Since the mid-1960s experience in this realm has revealed a third strand in the nature of the computer. Between the mathematics that makes the device theoretically possible and the electronics that makes it practically feasible lies the programming that makes it intellectually, economically, and socially useful. Unlike the extremes, the middle remains a craft, technical rather than technological, mathematical only in appearance. It poses the question of the relation of science and technology in a very special form.

That tripartite structure shows up in the three distinct disciplines that are concerned with the computer: electrical engineering, computer science, and software engineering. Of these, the first is the most well established, since it predates the computer, even though its current focus on microelectronics reflects its basic orientation toward the device. Computer science began to take shape during the 1960s, as it brought together common concerns from mathematical logic (automata, proof theory, recursive function theory), mathematical linguistics, and numerical analysis (algorithms, computational complexity), adding to them questions of the organization of information (data structures) and the relation of computer architecture to patterns of computation. Software engineering, conceived as a deliberately provocative term in 1967, has developed more as a set of techniques than as a body of learning. Except for a few university centers, such as Carnegie Mellon University, University of North Carolina, Berkeley, and Oxford, it remains primarily a concern of military and industrial R&D aimed at the design and implementation of large, complex systems, and the driving forces are cost and reliability.[13]

History of Computing as History of Technology

Consider, then, the history of computing in light of current history of technology. Several lines of inquiry seem particularly promising. Studies such as those cited above offer a panoply of models for tracing the patterns of growth and progress in computing as a technology. It is worth asking, for example, whether the computing industry has moved forward more by big advances of radical innovation or by small steps of improvement. Has it followed the process described by Nathan Rosenberg, whereby "technological improvement not only enters the structure of the economy through the main entrance, as when it takes the highly visible form of major patentable technological breakthroughs, but that it also employs numerous and less visible side and rear entrances where its arrival is unobtrusive, unannounced, unobserved, and uncelebrated"? To determine whether that is the case will require changes in the history of computing as it is currently practiced. It will mean looking beyond "firsts" to the revisions and modifications that made products work and that account for their real impact. Given the corporate, collaborative structure of modern R&D, historians of computing must follow the admonition once made to historians of technology to stop "substituting biography for careful analysis of social processes." Without denigrating the role of heroes and pioneers, we need more knowledge of computing's equivalent of "shop practices, [and of] the activities of lower-level technicians in factories." The question is how to pursue that inquiry across the variegated range of the emerging industry.[14]

Viewing computing both as a system in itself and as a component of a variety of larger systems may provide important insights into the dynamics of its development and may help to distinguish between its internal and its external history. For example, it suggests an approach to the question of the relation between hardware and software, often couched in the antagonistic form of one driving the other, a form which seems to assume that the two are relatively independent of one another. By contrast, linking them in a system emphasizes their mutual dependence. One expects of a system that the relationship among its internal components and their relationships to external components will vary over time and place but that they will do so in a way that maintains a certain equilibrium or homeostasis, even as the system itself evolves. Seen in that light, the relation between hardware and software is a question not so much of driving forces, or of stimulus and response, as of constraints and degrees of freedom. While in principle all computers have the same capacities as universal Turing machines, in practice different architectures are conducive to different forms of computing. Certain architectures have technical thresholds (e.g., VSLI is a prerequisite to massively parallel computing), others reflect conscious choices among equally feasible alternatives; some have been influenced by the needs and concerns of software production, others by the special purposes of customers. Early on, programming had to conform to the narrow limits of speed and memory set by vacuum-tube circuitry. As largely exogenous factors in the electronics industry made it possible to expand those limits, and at the same time drastically lowered the cost of hardware, programming could take practical advantage of research into programming languages and compilers. Researchers' ideas of multiuser systems, interactive programming, or virtual memory required advances in hardware at the same time that they drew out the full power of a new generation of machines. Just as new architectures have challenged established forms of programming, so too theoretical advances in computation and artificial intelligence have suggested new ways of organizing processors.[15]

At present, the evolution of computing as a system and of its interfaces with other systems of thought and action has yet to be traced. Indeed, it is not clear how many identifiable systems constitute computing itself, given the diverse contexts in which it has developed. We speak of the computer industry as if it were a monolith rather than a network of interdependent industries with separate interests and concerns. In addition to historically more analytical studies of individual firms, both large and small, we need analyses of their interaction and interdependence. The same holds for government and academia, neither of which has spoken with one voice on matters of computing. Of particular interest here may be the system-building role of the computer in forging new links of interdependence among universities, government, and industry after World War II.

Arguing in "The Big Questions" that creators of the machinery underpinning the American System worked from a knowledge of the entire sequence of operations in production, Daniels pointed to Peter Drucker's suggestion that "the organization of work be used as a unifying concept in the history of technology."[16] The recent volume by Charles Bashe et al. on *IBM's Early Computers* illustrates the potential fruitfulness of that suggestion for the history of computing. In tracing IBM's adaptation to the computer, they bring out the corporate tensions and adjustments introduced into IBM by the need to keep abreast of fast-breaking developments in science and technology and in turn to share its research with others.[17] The computer reshaped R&D at IBM, defining new relations between marketing and research, introducing a new breed of scientific personnel with new ways of doing things, and creating new roles, in particular that of the programmer. Whether the same holds true of, say, Bell Laboratories or G.E. Research Laboratories, remains to be studied, as does the structure of the R&D institutions established by the many new firms that constituted the growing computer industry of the 1950s, 1960s, and 1970s. Tracy Kidder's frankly journalistic account of development at Data General has given us a tantalizing glimpse of the patterns we may find. Equally important will be studies of the emergence of the data-processing shop, whether as an independent computer service or as a new element in established institutions.[18] More than one company found that the computer reorganized de facto the lines of effective managerial power.

The computer seems an obvious place to look for insight into the question of whether new technologies respond to need or create it. Clearly, the first computers responded to the felt need for high-speed, automatic calculation, and that remained the justification for their early development during the late 1940s. Indeed, the numerical analysts evidently considered the computer to be their baby and resented its adoption by "computerologists" in the late 1950s and early 1960s. But it seems equally clear that the computer became the core of an emergent data-processing industry more by creating demand than by responding to it. Much as Henry Ford taught the nation how to use an automobile, IBM and its competitors taught the nation's businesses (and its government) how to use the computer. How much of the technical development of the computer originated in the marketing division remains an untold story central to an understanding of modern technology. Kidder's *Soul of a New Machine* again offers a glimpse of what that story may reveal.[19]

One major factor in the creation of demand seems to have been the alliance between the computer and the nascent field of operations research/ management science. As the pages of the *Harvard Business Review* for 1953 show, the computer and operations research hit the business stage together, each a new and untried tool of management, both clothed in the mantle of science. Against the fanciful backdrop of Croesus' defeat by camel-riding Persians, an IBM advertisement proclaimed that "Yesterday . . . 'The Fates' Decided. Today . . . Facts Are What Count." Appealing to fact-based strides in "military science, pure science, commerce, and industry," the advertisement pointed beyond data processing to "'mathematical models' of specific processes, products, or situations, [by which] man today can predetermine probable results, minimize risks and costs." In less vivid terms, Cyril C. Herrmann of MIT and John F. Magee of Arthur D. Little introduced readers of *HBR* to "'Operations Research' for Management," and John Diebold proclaimed "Automation—The New Technology." As Herbert Simon later pointed out, operations research was both old and new, with roots going back to Charles Babbage and Frederick W. Taylor. Its novelty lay precisely in its claim to provide "mathematical models" of business operations as a basis for rational decision making. Depending for their sensitivity on computationally intensive algorithms and large volumes of data, those models required the power of the computer.[20]

It seems crucial for the development of the computer industry that the business community accepted the joint claims of OR and the computer long before either could validate them by, say, cost-benefit analysis. The decision to adopt the new methods of "rational decision making" seems itself to have been less than fully rational:

> As business managers we are revolutionizing the procedures of our factories and offices with automation, but what about out decision making? In other words, isn't there a danger that our thought processes will be left in the horse-and-buggy stage while our operations are being run in the age of nucleonics, electronics, and jet propulsion? . . . Are the engineering and scientific symbols of our age significant indicators of a need for change?[21]

Even at this early stage, the computer had acquired symbolic force in the business community and in society at large. We need to know the sources of that force and how it worked to weave the computer into the economic and social fabric.[22]

The government has played a determining role in at least four areas of computing: microelectronics; interactive, real-time systems; artificial intelligence; and software engineering. None of these stories has been told by a historian, although each promises deep insight into the issues raised above. Modern weapons systems and the space program placed a premium on miniaturization of circuits. Given the costs of research, development, and tooling for production, it is hard to imagine that the integrated circuit and the microprocessor would have emerged—at least as quickly as they did—without government support. As Frank Rose put it in *Into the Heart of the Mind*, "The computerization of society . . . has essentially been a side effect of the computerization of war." More is involved than smaller computers. Architecture and software change in response to speed of processor and size of memory. As a result, the rapid pace of miniaturization tended to place already inadequate methods of software production under the pressure of rising expectations. By the early 1970s the Department of Defense, as the nation's single largest procurer of software, had declared a major stake in the development of software engineering as a body of methods and tools for reducing the costs and increasing the reliability of large programs.[23]

As Howard Rheingold has described in *Tools for Thought*, the government was quick to seize on the interest of computer scientists at MIT in developing the computer as an enhancement and extension of human intellectual capabilities. In general, that interest coincided with the needs of national defense in the form of interactive computing, visual displays of both text and graphics, multiuser systems, and inter-computer networks. The Advanced Research Projects Agency (later DARPA), soon became a source of almost unlimited funding for research in these areas, a source that bypassed the usual procedures of scientific funding, in particular peer review. Much of the early research in artificial intelligence derived its funding from the same source, and its development as a field of computer science surely reflects that independence from the agenda of the discipline as a whole.[24]

Although we commonly speak of hardware and software in tandem, it is worth noting that in a strict sense the notion of software is an artifact of computing in the business and government sectors during the 1950s. Only when the computer left the research laboratory and the hands of the scientists and engineers did the writing of programs become a question of production. It is in that light that we may most fruitfully view the development of programming languages, programming systems, operating systems, database and file management systems, and communications and networks, all of them aimed at facilitating the work of programmers, maintaining managerial control over them, and assuring the reliability of their programs. The Babel of programming languages in the 1960s tends to distract attention from the fact that three of the most commonly used languages today are also among the oldest: FORTRAN for scientific computing, COBOL for data processing, and LISP for artificial intelligence. ALGOL might have remained a laboratory language had it and its offspring not become the vehicles of structured programming, a movement addressed directly to the problems of programming as a form of production.[25]

Central to the history of software is the sense of "crisis" that emerged in the late 1960s as one large project after another ran behind schedule, over budget, and below specifications. Though pervasive throughout the industry, it posed enough of a strategic threat for the NATO Science Committee to convene an international conference in 1968 to address it. To emphasize the need for a concerted effort along new lines, the committee coined the term "software engineering," reflecting the view that the problem required the combination of science and management thought characteristic of engineering. Efforts to define that combination and to develop the corresponding methods constitute much of the history of computing during the 1970s, at least in the realm of large systems, and it is the essential background to the story of Ada in the 1980s. It also reveals apparently fundamental differences between the formal, mathematical orientation of European computer scientists and the practical, industrial focus of their American counterparts. Historians of science and technology have seen those differences in the past and have sought to explain them. Can historians of computing use those explanations and in turn help to articulate them?

The effort to give meaning to "software engineering" as a discipline and to define a place for it in the training of computer professionals should call the historian's attention to the constellation of questions contained under the heading of "discipline formation and professionalization." In 1950 computing consisted of a handful of specially designed machines and a handful of specially trained programmers. By 1955 some 1,000 general-purpose computers required the services of some 10,000 programmers. By 1960, the number of devices had increased fivefold, the number of programmers sixfold. And so the growth continued. With it came associations, societies, journals, magazines, and claims to professional and academic standing. The development of these institutions is an essential part of the social history of computing as a technological enterprise. Again, one may ask to what extent that development has followed historical patterns of institutionalization and to what extent it has created its own.

The question of sources illustrates particularly well how recent work in the history of technology may provide important guidance to the history of computing, at the same time that the latter adds new perspectives to that work. As noted above, historians of technology have focused new attention on the non-verbal expressions of engineering practice. Of the three main strands of computing, only theoretical computer science is essentially verbal in nature. Its sources come in the form most familiar to historians of science, namely books, articles, and other less formal pieces of writing, which by and large encompass the thinking behind them. We know pretty well how to read them, even for what they do not say explicitly. Similarly, at the level of institutional and social history, we seem to be on familiar ground, suffering largely from an embarrassment of wealth unwinnowed by time.

But the computers themselves and the programs that were written for them constitute a quite different range of sources and thus pose the challenge of determining how to read them. As artifacts, computers present the problem of all electrical and electronic devices. They are machines without moving parts. Even when they are running, they display no internal action to explain their outward behavior. Yet, Tracy Kidder's portrait of Tom West sneaking a look at the boards of the new Vax to see how DEC had gone about its work reminds us that the actual machines may hold tales untold by manuals, technical reports, and engineering drawings. Those sources too demand our attention. When imaginatively read, they promise to throw light not only on the designers but also on those for whom they were designing. Through the hardware and its attendant sources one can follow the changing physiognomy of computers as they made their way from the laboratories and large installations to the office and the home. Today's prototypical computer iconically links television to typewriter. How that form emerged from a roomful of tubes and switches is a matter of both technical and cultural history.[26]

Though hard to interpret, the hardware is at least tangible. Software by contrast is elusively intangible. In essence, it is the behavior of the machines when running. It is what converts their architecture to action, and it is constructed with action in mind; the programmer aims to make something happen. What, then, captures software for the historical record? How do we document and preserve a historically significant compiler, operating system, or database? Computer scientists have pointed to the limitations of the static program text as a basis for determining the program's dynamic behavior, and a provocative article has questioned how much the written record of programming can tell us about the behavior of programmers. Yet, Gerald M. Weinberg has given an example of how programs may be read to reveal the machines and people behind them. In a sense, historians of computing encounter from the opposite direction the problem faced by the software industry: what constitutes an adequate and reliable surrogate for an actually running program? How, in particular, does the historian recapture, or the producer anticipate, the component that is always missing from the static record of software, namely, the user for whom it is written and whose behavior is an essential part of it?[27]

Placing the history of computing in the context of the history of technology promises a peculiarly recursive benefit. Although computation by machines has a long history, computing in the sense I have been using here did not exist before the late 1940s. There were no computers, no programmers, no computer scientists, no computer managers. Hence those who invented and improved the computer, those who determined how to program it, those who defined its scientific foundations, those who established it as an industry in itself and introduced it into business and industry, all came to computing from some other background. With no inherent precedents for their work, they had to find their own precedents. Much of the history of computing, certainly for the first generation, but probably also for the second and third, derives from the precedents these people drew from their past experience. In that sense, the history of technology shaped the history of computing, and the history of computing must turn to the history of technology for initial bearings.

A specific example may help to illustrate the point. Daniels stated as one of the really big questions the development of the "American System" and its culmination in mass production. It is perhaps the central fact of technology in nineteenth-century America, and every historian of the subject must grapple with it. So too, though Daniels did not make the point, must historians of twentieth-century technology. For mass production has become a historical touchstone for modern engineers, in the area of software as well as elsewhere.[28]

For instance, in one of the major invited papers at the NATO Software Engineering Conference of 1968, M. D. McIlroy of Bell Telephone Laboratories looked forward to the end of a "preindustrial era" in programming. His metaphors and similes harked back to the machine-tool industry and its methods of production.

> We undoubtedly produce software by backward techniques. We undoubtedly get the short end of the stick in confrontations with hardware people because they are the industrialists and we are the crofters. Software production today appears in the scale of industrialization somewhere below the more backward construction industries. I think its proper place is considerably higher, and would like to investigate the prospects for mass-production techniques in software.[29]

What McIlroy had in mind was not replication in large numbers, which is trivial for the computer, but rather programmed modules that might serve as standardized, interchangeable parts to be drawn from the library shelf and inserted in larger production programs. A quotation from McIlroy's paper served as leitmotif to the first part of Peter Wegner's series "Capital Intensive Software Technology" in the July 1984 number of *IEEE Software*, which was richly illustrated by photographs of capital industry in the 1930s and included insets on the history of technology. By then

McIlroy's equivalent to interchangeable parts had become "reusable software" and software engineers had developed more sophisticated tools for producing it. Whether they were (or now are) any closer to the goal is less important to the historian than the continuing strength of the model. It reveals historical self-consciousness.[30]

We should appreciate that self-consciousness at the same time that we view it critically, resisting the temptation to accept the comparisons as valid. An activity's choice of historical models is itself part of the history of the activity. McIlroy was not describing the state or even the direction of software in 1968. Rather, he was proposing a historical precedent on which to base its future development. What is of interest to the historian of computing is why McIlroy chose the model of mass production as that precedent. Precisely what model of mass production did he have in mind, why did he think it appropriate or applicable to software, why did he think his audience would respond well to the proposal, and so on? The history of technology provides a critical context for evaluating the answers, indeed for shaping the questions. For historians, too, the evolving techniques of mass production in the nineteenth century constitute a model, or prototype, of technological development. Whether it is one model or a set of closely related models is a matter of current scholarly debate, but some features seem clear. As a system it rested on foundations established in the early and mid-nineteenth century, among them in particular the development of the machine-tool industry, which, as Nathan Rosenberg has shown, itself followed a characteristic and revealing pattern of innovation and diffusion of new techniques. Even with the requisite precision machinery, methods of mass production did not transfer directly or easily from one industry to another, and its introduction often took place in stages peculiar to the production process involved. Software production may prove to be the latest variation of the model, or critical history of technology may show how it has not fit.[31]

Conclusion: The Real Computer Revolution

We can take this example a step farther. From various perspectives, people have been drawn to compare the computer to the automobile. Apple, Atari, and others have boasted of creating the Model T of microcomputers, clearly intending to convey the image of a car in every garage, an automobile that everyone could drive, a machine that reshaped American life. The software engineers who invoke the image of mass production have it inseparably linked in their minds to the automobile and its interchangeable variations on a standard theme.

The two analogies serve different aims within the computer industry, the first looking to the microcomputer as an object of mass consumption, the second to software systems as objects of mass production. But they share the vision of a society radically altered by a new technology. Beneath the comparison lies the conviction that the computer is bringing about a revolution as profound as that triggered by the automobile. The comparison between the machines is fascinating in itself. Just how does one weigh the PC against the PT (personal transporter)?[32] For that matter, which PC is the Model T: the Apple II, the IBM, the Atari ST, the Macintosh? Yet the question is deeper than that. What would it mean for a microcomputer to play the role of the Model T in determining new social, economic, and political patterns? The historical term in that comparison is not the Model T, but Middletown, where in less than forty years "high-speed steel and Ford cars" had fundamentally changed the nature of work and the lives of the workers.[33] Where is the Middletown of today, similarly transformed by the presence of the microcomputer? Where would one look? How would one identify the changes? What patterns of social and intellectual behavior mark such transformation? In short, how does one compare technological societies? That is one of the "big questions" for historians of technology, and it is only in the context of the history of technology that it will be answered for the computer.

From the very beginning, the computer has borne the label "revolutionary." Even as the first commercial machines were being delivered, commentators were extolling or fretting over the radical changes the widespread use of computers would entail, and few doubted their use would be widespread. The computer directed people's eyes toward the future, and a few thousand bytes of memory seemed space enough for the solution of almost any problem. On that both enthusiasts and critics could agree. Computing meant unprecedented power for science, industry, and business, and with the power came difficulties and dangers that seemed equally unprecedented. By its nature as well as by its youth, the computer appeared to have no history.

Yet, "revolution" is an essentially historical concept.[34] Even when turning things on their head, one can only define what is new by what is old, and innovation, however imaginative, can only proceed from what exists. The computer had a history out of which it emerged as a new device, and computing took shape from other, continuing activities, each with its own historical momentum. As the world of the computer acquired its own form, it remained embedded in the worlds of science, technology, industry, and business, which structured computing even as they changed in response to it. In doing so they linked the history of computing to their own histories, which in turn reflected the presence of a fundamentally new resource.

What is truly revolutionary about the computer will become clear only when computing acquires a proper history, one that ties it to other technologies and thus uncovers the precedents that make its innovations significant. Pursued within the larger enterprise of the history of technology, the history of computing will acquire the context of place and time that gives history meaning.

CHAPTER TWO

What Makes History?

A s you look over the reviewers' and Program Committee's comments on your paper, you may be perplexed by admonitions in one form or the other to "make it more historical." After all, you've been talking about the past, you've got the events and people in chronological order, you've related what happened. What more could be needed to make history? The answer is hard to pin down in a series of methodological precepts, as history is ultimately an art acquired by professional practice. But it may help you to understand our specific criticisms if I describe what we're looking for in general terms.

Dick Hamming captured the essence of it in the title of his paper at the International Conference on the History of Computing held in Los Alamos in 1976, "We Would Know What They Thought When They Did It."[1] He pleaded for a history of computing that pursued the contextual development of ideas, rather than merely listing names, dates, and places of "firsts." Moreover, he exhorted historians to go beyond the documents to "informed speculation" about the results of undocumented practice. What people actually did and what they thought they were doing may well not be accurately reflected in what they wrote and what they said they were thinking. His own experience had taught him that.

Getting behind the documentation to discover what people were thinking is no easy task, even when the people are still available to ask or when they themselves are doing the history. A story, perhaps apocryphal, about Jean Piaget shows why. The psychologist was standing outside one evening with a group of eleven-year-olds and called their attention to the newly risen moon, pointing out that it was appreciably higher in the sky than it had been at the same time the night before and wondering why that was. The children were also puzzled, though in their case genuinely so. In his practiced way, Piaget led them to discover the relative motions of the earth, moon, and sun and thus to arrive at an explanation. A month or two later, the same group was together under similar circumstances, and Piaget again posed his question. "That's easy to explain," said one boy, who proceeded to sketch out the motions that accounted for the phenomenon. "That's remarkable," said Piaget. "How did you know that?" "Oh," the boy replied, "we've always known that!"

Not only children, but people in general, and scientists in particular, quickly forget what it was like not to know what they now know. That is, once you've solved a problem, especially when the solution involved a new approach, it's difficult to think about the problem in the old way. What was once unknown has become obvious. What once tested the ingenuity of the skilled practitioner is now "an exercise left to the student." The phenomenon affects not only the immediate past, but the distant past as well. When scientists study history, they often use their modern tools to determine what past work was "really about"; for example, the Babylonian mathematicians were "really" writing algorithms. But that's precisely what was not "really" happening. What was really happening was what was possible, indeed imaginable, in the intellectual environment of the time; what was really happening was what the linguistic and conceptual framework then would allow. The framework of Babylonian mathematics had no place for a metamathematical notion such as algorithm.

These considerations suggest the paradox inherent in the role of pioneers and participants in the history of computing, whether as active members of panels and workshops, as subjects of interviews, or as historians of their own work. On the one hand, they are our main source for knowing "what they thought when they did it," not only for their own efforts, but also those of their colleagues and contemporaries. On the other, it is they who reshaped our understanding by their discoveries, solutions, and inventions and who, as a result, may find it harder than most to recall just what they were thinking. Precisely because they helped to create the present, they are prone to identify it with their past work or to translate that work into current terminology.

Doing so defeats the utility, or even the purpose, of their testimony. To the historian, the old way is crucial: it holds the roots of the new way. It does not illuminate history to say "We were really doing X," where X stands for the current state of the art. Talking that way masks the very changes in conceptual structure that explain the development of X and that history aims at elucidating. It may well be that the roots of a modern technique or theory lie in work done thirty or forty years ago. That can only become clear by tracing the growth of the tree, determining its branching pattern, and identifying the points at which grafting has occurred. That requires in turn that the root be characterized in its own terms. When a pioneer talks about his or her work in the past, it is important for others to listen critically and be ready to say, "Now wait a second, we didn't put the question that way at the time, nor did we know that concept." It is important for the pioneer to develop the same sensitivity to changes in language over time. They point to changes in conceptualization. Historically, a rose by another name may have a quite different smell.

Having the solution can mask the original problem in several ways. One may forget there was a problem at all, or undervalue the urgency it had, projecting its current insignificance back to a time when it was not trifling at all, but rather a serious concern. One may reconstruct a different problem, overlooking the restructuring of subject brought about by the solution. One may ignore or undervalue alternative solutions that once looked attractive and could very well have taken development in a different direction. Historically, a "right" answer requires just as much explanation as a "wrong" answer, and both answers are equally interesting—and equally important.

Several years ago I asked someone at Bell Labs responsible for maintaining software what sort of documentation she would like most to have but did not. "I'd like to know why people didn't do things," she said. "In many cases when a problem arises, we look at the program and think we see another, better way of doing things. We spend six months pursuing that alternative only to discover that it won't work, and we then realize why the original team didn't choose it. We'd like to know that before we start." Something quite close to that holds for historians too. They want to know what the choices and possibilities were at a given time, why a particular one was adopted, and why the others were not. Good history, from the historian's point of view, always keeps the options in view at any given time.

Getting the facts right is important, both the technical facts and the chronological facts. But the reasons for those facts are even more important, and the reasons often go well beyond the facts. When people come together on a project or for a meeting—when, indeed, they are creating a new enterprise—they bring with them their past experience and their current concerns, both as individuals and as members of institutions. In both cases, they have interests and commitments that transcend the problem at hand, yet determine its shape and the range of acceptable solutions. It is essential to know, in both the real and the idiomatic sense of the phrase, where they are coming from. In some cases, discretion, proprietary information, or just plain ignorance may preclude a definitive answer. Nonetheless, even when we can't know the answers, it is important to see the questions. They too form part of our understanding. If you cannot answer them now, you can alert future historians to them.

What makes history? It's a matter of going back to the sources, of reading them in their own language, and of thinking one's way back into the problems and solutions as they looked then. It involves the imaginative exercise of suspending one's knowledge of how things turned out so as to re-create the possibilities still open at the time. In *The Go-Between*, Leslie Hartley remarked that "The past is a foreign country; they do things differently there." The historian must learn to be an observant tourist, alert to the differences that lie behind what seems familiar.

CHAPTER THREE

Issues in the History of Computing

I T SHOULD BE EASY to do the history of computing. After all, computing began less than fifty years ago, and we have crowds of eyewitnesses, mountains of documents, storerooms of original equipment, and the computer itself to process all the information those sources generate. What's so hard? What are we missing?

Well, the record is not quite as complete as it looks. As the Patent Office discovered during the 1980s, much of the art of programming is undocumented, as is much of the software created during the 1950s and since. The software itself is becoming inaccessible, as the machines on which it ran disappear through obsolescence. Critical decisions lie buried in corporate records—either literally so or as trees are in a forest. Many eyewitnesses have testified, but few have been cross-examined. Much of the current record exists only online and, even, if archived will consist of needles in huge haystacks. But these are minor matters, to be discussed in the context of broader issues.

The major problem is that we have lots of answers but very few questions, lots of stories but no history, lots of things to do but no sense of how to do them or in what order. Simply put, we don't yet know what the history of computing is really about. A glance at the literature makes that clear. We still know more about the calculating devices that preceded the electronic digital computer—however tenuously related to it—than we do about the machines that shaped the industry. We have hardly begun to determine how the industry got started and how it developed. We still find it easier to talk about hardware than about software, despite the shared sense that much of the history of computing is somehow wrapped up in the famous inversion curve of hardware/software costs from 1960 to 1990. We still cast about for historical precedents and comparisons, unsure of where computing fits into the society that created it and has been shaped by it. That uncertainty is reflected in the literature and activities of the History of Science Society and the Society for the History of Technology, where the computer as the foremost scientific instrument, and computing as the salient technology, of the late twentieth century are all but invisible, not only in themselves but in the perspective they shed on earlier science and technology.

The presentations in the second session today are addressed to the various materials that constitute the primary sources for the history of computing: artifacts, archives and documents, first-person experience both written and oral. The business of HOPL-II itself is to enrich the sources for programming languages. As I noted at the outset, we seem to have no shortage of such materials. Indeed, as someone originally trained as a medievalist, I feel sometimes like a beggar at a banquet. I hardly know where to begin, and my appetite runs ahead of my digestion. It's a real problem, and not only for the consumer. To continue the metaphor, at some point the table can't hold all the dishes, and the pantry begins to overflow.

We have to pick and choose what we keep in store and what we set out. But by what criteria? Conflagration has done a lot of the selecting for medievalists. Historians of computing—indeed of modern science and technology in general—have to establish principles of a less random character. Once collecting and preserving become selective (I mean consciously selective; they are always unconsciously selective), they anticipate the history they are supposed to generate and thus create that history. That is, they are historical activities, reflecting what we think is important about the present through what we preserve of it for the future. What we think is important about the present depends on who we are, where we stand—in the profound sense of the cliché, where we're coming from. Everyone at HOPL-II is doing history simply by being here and ratifying through our presence that the languages we are talking about and what we say about their origin and development are historically significant.

There's an important point here. Let me bring it out by means of something that seems to be of consuming interest to computer people, namely, "firsts." Who was first to . . . ? What was the first . . . ? Now, that can be a tricky question because it can come down to a matter of meaning rather than of order in time. Nothing is entirely new, especially in matters of scientific technology. Innovation is incremental, and what already exists determines to a large extent what can be created. So collecting and recording "firsts" means deciding what makes them first, and that decision can often lead to retrospective judgments, where the first X had always been known as a Y. Let me give an example prompted by a distinguished computer scientist's use of history. I want to return to the topic at the end of this paper, so the example is more detailed than it need be for the present point.

In a review of business data processing in 1962, Grace Hopper, then at Remington Rand UNIVAC Division, sought to place the computer in the mainstream of American industrial development by emphasizing a characteristic sequence of venture—enterprise—mass production—adventure. The model suggested that computing was entering a period of standardization before embarking in daring new directions. The first two stages are tentative and experimental, but

> as the enterprise settles down and becomes practical mass production, the number of models diminishes while the number made of each model increases. This is the process, so familiar to all of us, by which we have developed our refrigerators, our automobiles, and our television sets. Sometimes we forget that in each case the practical production is followed by the new adventure. Sometimes we forget that the Model T Ford was followed by one with a gear shift.[1]

That Model T, an icon of American industrialism, deserves a closer look. Shifting-gear transmissions antedated the Model T. But the softness of the steel then available and the difficulty of meshing gears meant the constant risk of stripping them. Concerned above all with reliability and durability, Ford consciously avoided shifting by going back to the planetary transmission, in which the gears remain enmeshed at all times and are brought to bear on the driveshaft in varying combinations by bands actuated by foot pedals. Although the Model A indeed had a gearshift, made reliable by Ford's use of new alloys, it is perhaps more important historically that the planetary transmission had just begun its life. For later, with the foot pedals replaced by a hydraulic torque converter, it served as the heart of the automatic transmissions that came to dominate automotive design in the 1950s. Now, how does one unravel the "firsts" in all this? Would we think to keep a Model T around for the history of automatic transmissions? What then of the first operating system, or the first database? Does the equivalent of a planetary transmission lurk in them, too?

Documenting Practice

In deciding what to keep, it may help to understand that historians look at sources not only for what is new and unusual but also for what is so common as to be taken for granted. In the terms of information theory, they are equally interested in the redundancy that assures the integrity of the message. For much of common practice is undocumented, and yet it shapes what is documented. The deep effect of innovation is to change what we take for granted. This is what Thomas S. Kuhn had in mind with the notion of "paradigm shift," the central concept of his immensely influential book, *The Structure of Scientific Revolutions*. He later replaced "paradigm" with "disciplinary matrix," but the meaning remained the same: in learning to do science, we must learn much more than is in the textbook. In principle, knowing that $F = ma$ is all you need to know to solve problems in classical mechanics. In practice, you need to know a lot more than that. There are tricks to applying $F = ma$ to particular mechanical systems, and you learn how to do it by actually solving problems under the eye of someone who already has the skill. It is that skill, a body of techniques and the habits of thought that go with them, that constitutes effective knowledge of a subject.[2]

Because all practitioners of a subject share that skill, they do not talk about it. They take it for granted. Yet it informs their thinking at the most basic level, setting the terms in which they think about problems and shaping even the most innovative solutions. Thus it plays a major role in deciding about "firsts." Over time, the body of established practice changes, as new ideas and techniques become common knowledge and older skills become obsolete. Again, one doesn't talk much about it; "everyone knows that!" Yet, it is fragile. It is hard to keep, much harder than machines or programs.

To gain access to undocumented practice, historians are learning to do what engineers often take for granted, to read the products of practice in critical ways. In *The Soul of a New Machine*, Tracy Kidder relates how Tom West of Data General bluffed his way into a company that was installing a new VAX and spent the morning pulling boards and examining them.

> Looking into the VAX, West had imagined he saw a diagram of DEC's corporate organization. He felt that the VAX was too complicated. He did not like, for instance, the system by which various parts of the machine communicated with each other; for his taste, there was too much protocol involved. He decided that VAX embodied flaws in DEC's corporate organization. The machine expressed that phenomenally successful company's cautious, bureaucratic style. Was this true? West said it didn't matter, it was a useful theory.[3]

Historically, it is indeed a useful theory. Technology is not a literate enterprise; not because inventors and engineers are illiterate, but because they think in things rather than words. Henry Ford summed it up in his characteristically terse style:

> There is an immense amount to be learned simply by tinkering with things. It is not possible to learn from books how everything is made—and a real mechanic ought to know how nearly everything is made. Machines are to a mechanic what books are to a writer. He gets ideas from them, and if he has any brains he will apply those ideas.[4]

In short, the record of technology lies more in the artifacts than in the written records, and historians have to learn to read the artifacts as critically as they do the records. It is perhaps the best way of meeting Dick Hamming's challenge to "know what they thought when they did it." For that, historians need, first and foremost, the artifacts. That is where museums play a central role in historical research. But historians also need help in learning how to read those artifacts. That means getting the people who designed and built them to talk about the knowledge and know-how that seldom gets into words.[5]

As I use the word "artifact" here, I suspect that most of the audience has a machine in mind. But computing has another sort of artifact, one that seems unique until one thinks about it carefully. I mean a program: a high-level language compiler, an operating system, an application. It has been the common lament of management that programs are built by tinkering and that little of their design gets captured in written form, at least in a written form that would make it possible to determine how they work or why they work as they do rather than in other readily imaginable ways. Moreover, what programs do and what the documentation says they do are not always the same thing. Here, in a very real sense, the historian inherits the problems of software maintenance: the farther the program lies from its creators, the more difficult it is to discern its architecture and the design decisions that inform it.

Yet software can be read. In Alan Kay's talk, we'll hear a counterpart to Tom West's reading of the VAX boards:

> Head whirling, I found my desk. On it was a pile of tapes and listings, and a note: "This is the ALGOL for the 1108. It doesn't work. Please make it work". The latest graduate student gets the latest dirty task.
>
> The documentation was incomprehensible. Supposedly, this was the Case-Western Reserve 1107 ALGOL—but it had been doctored to make a language called Simula; the documentation read like Norwegian transliterated into English, which in fact was what it was. There were uses of words like *activity* and *process* that didn't seem to coincide with normal English usage.
>
> Finally, another graduate student and I unrolled the listing 80 feet down the hall and crawled over it yelling discoveries to each other. The weirdest part was the storage allocator, which did not obey a stack discipline as was

usual for ALGOL. A few days later, that provided the clue. What Simula was allocating were structures very much like instances of Sketchpad.[6]

In the draft version, Kay added as a gloss to his tale that such explorations of machine code were common among programmers: "Batch processing and debugging facilities were so bad back then that one would avoid running code at any cost. 'Desk-checking' listings was the way of life." Yet, it is not the sort of thing one finds in manuals or textbooks. Gerald M. Weinberg gives an example of how it is done for various versions of FORTRAN in Chapter 1 of *The Psychology of Computer Programming*, and Brian Kernighan and P. J. Plauger's *Elements of Programming Style* can be read as a guide to the art of reading programs. But the real trick is to capture the know-how, the tricks of the trade that everyone knew and no one wrote down when "desk-checking" was the norm. We need to think about how best to do that, because for historians without access to the machines on which programs ran or without the resources to have emulators written, desk-checking could again become a way of life. At the very least, it will demand fluency in the languages themselves.[7]

The Importance of Context

An emphasis on practice is also an emphasis on context. In focusing on what was new about computing in general and about various aspects of it in particular, one can lose sight of other elements of tradition (in the literal sense of "handing over") and of training that determined what was taken for granted as the basis for innovation. From the outset, computing was what Derek J. de Solla Price, a historian and sociologist of science, termed "big science." It relied on government funding, an expanding economy, an advanced industrial base, and a network of scientific and technical institutions. Over the past forty-odd years it has achieved autonomy as a scientific and technical enterprise, but it did so on the basis of older, established institutions, from which the first generations of computer people brought their craft practices, their habits of thought, their paradigms, and their precedents, all of which shaped the new industry and discipline they were creating.[8]

As in many other things, Alan Perlis offered a productive metaphor for thinking about context. At the first HOPL he reflected on the fate of ALGOL 58 in competition with FORTRAN, noting that:

> The acceptance of FORTRAN within SHARE and the accompanying incredible growth in its use, in competition with a better linguistic vehicle, illustrated how factors other than language determine the choice of programming

languages of users. Programs are not abstractions. They are articles of commerce in a society of users and machines. The choice of programming language is but one parameter governing the vitality of the commerce.[9]

A comment by Kristen Nygaard during discussion led Perlis to expand the point:

> I think that my use of the word 'commerce' has been, at least in that case, misinterpreted. My use of the word 'commerce' was not meant to imply that it was IBM's self-interest which determined that FORTRAN would grow and that ALGOL would wither in the United States. It certainly was the case that IBM, being the dominant computer peddler in the United States, determined to a great extent—that the language FORTRAN would flourish, but IBM has peddled other languages which haven't flourished. FORTRAN flourished because it fitted well into the needs of the customers, and that defined the commerce. SHARE is the commerce. The publication of books on FORTRAN is the commerce. The fact that every computer manufacturer feels that if he makes a machine, FORTRAN must be on it is part of the commerce. In Europe, ALGOL is part of a commerce that is not supported by any single manufacturer, but there is an atmosphere in Europe of participation in ALGOL, dissemination of ALGOL, education in ALGOL, which makes it a commercial activity. That's what I mean by 'commerce'—the interplay and communication of programs and ideas about them. That's what makes commerce, and that what makes languages survive.[10]

One can extend Perlis's metaphor ever further, shifting the focus from survival to design. Henry Ford insisted that the Model T embodied a theory of business. He had designed it with a market in mind. That same may be said of any product, including programs. An industrial artifact is designed with a consumer in mind and hence reflects the designers' view of the consumer. The market—another term for "commerce"—is thus not a limiting condition, an external constraint on the product, but rather a defining condition built into the product. "What does this mean?" can often best be answered by determining for whom it was meant. For both hardware and software, that may not be an easy thing to do. Stated goals may not have coincided with unstated, the people involved may not have agreed, and the computing world has a talent for justifying itself in retrospect.

As Perlis suggested, "commerce" in a general sense extends beyond industry and business to encompass science, technology, and the institutions that support them. The notion emphasizes the role of institutions in directing the technical development of computing and, to some extent, conversely. In *Creating the Computer*, Kenneth Flamm has revealed the patterns of government support that gave the computer and computing their initial shape. The recent historical report by Arthur Norberg and Judy O'Neill on DARPA's Information Processing Techniques Office explores the interplay between defense needs and academic interests in the development of time-sharing, packet-switched networks, interactive computer graphics, and artificial intelligence. What is particularly striking is the mobility of personnel between government offices and university laboratories, making it difficult at times to discern just who is designing what for whom. The study of the NSF's program in computer science, now being completed by William Aspray and colleagues, opens similar insights into the reciprocal influences between programs of research and sources of funding. In *IBM's Early Computers*, Charles Bashe and his co-authors show how commercialization of a cutting-edge technology forced the corporation, until then used to self-reliance in research and development, to open new links with the larger technical community and to play an active role in it, as evidenced by the *IBM Journal of Research and Development*, first published in 1957. Closer to my own home, the development of computer science at Princeton has been characterized by the easy flow of researchers between the University and Bell Labs, and that same pattern has surely obtained elsewhere.[11]

Taken broadly, then, the notion of "commerce" directs the historian's attention to the determinative role of the marketplace in modern scientific technologies such as computing. It is one thing to build number crunchers one by one on contract to the government for its laboratories; it is another to develop a product line for a market that must be created and fostered. The explosive growth of computing since the early 1950s depended on the ability of the industry to persuade corporations and, later, individuals that they needed computers, or at least could benefit from them. That meant not only designing machines but also, and increasingly, uses for them in the form of computer systems and applications. Henry Ford put Americans on wheels in part by showing them how they could use an automobile. Similarly, the computing industry has had to create uses for the computer. The development of computing as a technology has depended, at least in part, on its success in doing so, and hence understanding the directions the technology has taken, even in its most scientific form, means finding the market for which it was being designed, or the market it was trying to design.

"Market" here includes people and people's skills. Earlier, I said a program seems "unique until one thinks about it." One may think about a program as essentially the design of a machine through the organization of a system of production. Other people thought about the organization of production and the management of organizations, and in many cases creating a market for computing meant creating a market for the skills of organizing systems of production. In examining the contexts of computing, historians would do well to explore, for example, the close relations between computing and industrial engineering and management. The *Harvard Business Review* introduced its readers to the computer and to operations research in back-to-back issues in 1953, and the two technologies have had a symbiotic relation ever since. Both had something to sell, both had to create a market for their products, and they needed each other to create it. Computers and computing have evolved in a variety of overlapping contexts, shaping those contexts while being shaped by them. The history of computing thus lies in the intersections of the histories of many technologies, to which the historian of computing must remain attuned.[12]

The notion of "computing as commerce" brings out the significance of techniques of reading artifacts, combined with the more usual techniques of textual criticism. To repeat: from the outset, computing has had to sell itself, whether to the government as big machines for scientific computing essential to national defense, to business and industry as systems vital to management, or to universities as scientific and technological disciplines deserving of academic standing and even departmental autonomy. The computing community very quickly learned the skills of advertising and became adept at marketing what it often could not yet produce. The result is that computing has had an air of wishful thinking about it. Much of its literature interweaves performance with promise, what is in practice with what can be in principle. It is a literature filled with announcements of revolutions subsequently (and quietly) canceled owing to unforeseen difficulties. In the case of confessed visionaries like Ted Nelson, the sources carry their own caveat. But in many instances computer marketers and management consultants, not to mention software engineers, were no less visionary, if perhaps less frank about it. What sources claimed or suggested could be done did not always correspond to what in fact could be done at the time. An industry trying to expand its market, engineers and scientists trying to establish new disciplines and attract research and development funding, new organizations seeking professional standing, did not talk a lot about failures. The artifacts, both hard and soft, are the firmest basis for separating fact from fiction technically, provided one learns to read them critically. Doing so is essential to understanding the claims made for computers and programs, and why they were believed.

Doing History

Recognizing the elements of continuity that link computing to its own past, and to the past of the industries and institutions that have fostered it, brings out most clearly the relation of history and current practice. History is built into current practice; the more profound the influence, the less conscious we are of its presence. It is not a matter of learning the "lessons of history" or of exploring "what history can teach us," as if these were alternatives or complements to practice. We are products of our history or, rather, our histories; we do what we do for historical reasons. Bjarne Stroustrup notes in his history of C++, "We never have a clean slate. Whatever new we do we must also make it possible for people to make a transition from old tools and ideas to new."[13]

History serves its purpose best, not when it suggests what we should be doing but when it makes clear what we *are* doing or at least clarifies what we *think* we are doing. On matters that count, we invoke precedents, which is to say we invoke history. We should be conscious of doing that, and we should be concerned both that we have the right history and that we have the history right.

Earlier, to make a point about "firsts," I cited Grace Hopper's evocation of the Model T. You may well have felt that my critique focused on a matter of detail that is irrelevant to her main point. Perhaps; but consider the historical basis of that main point. She was taking the automobile industry as a precedent for business data processing and by extension for computing as a whole. It was not the first time she had done so. The precedent she invoked in her famous "Education of a Computer" was also taken from the automotive production line, and it has persisted as a precedent down to the present: look at the cover of *IEEE Software* for June 1987, where the Ford assembly line around 1940 serves as backdrop to Peter Wegner's four-part article on industrial-strength software; then look at the Ford assembly line of the 1950s on the jacket of Greg Jones's *Software Engineering* in 1990. These are not isolated examples, nor are they mere window dressing. They reflect a way of thinking about software engineering, and there is history built into it, as there is in the notion of engineering itself; witness Mary Shaw's comparison of software engineering with other engineering disciplines. There were other ways to think about writing programs for computers. If people thought, and even continue to think, predominantly about automobiles or engineering, it is for historical reasons.[14]

In commenting on a draft of this paper, Bob Rosin noted at this point that "automobiles and engineering are (outdated?) paradigms that are largely unfamiliar to younger computer people. Many kids, who grew up hacking on Atari's and PC's and Mac's, didn't hack automobiles and rarely studied engineering." The implied objection touches my argument only where I shift from history to criticism of current practice. Historically, at least through the 1980s, software engineering has taken shape with reference to the assembly line and to industrial engineering as models. Those models are built into the current enterprise. One only has to read Doug McIlroy's "On Mass-Produced Software Components," presented to the first NATO Software Engineering Conference in 1968, to see where the conceptual roots of object-oriented programming lie. Ignorance of the automobile and engineering, or at least of their role in the formation of software engineering, will not free a new generation of software developers from the continuing influence of the older models built into the practice they learn and the tools they use. Rather, it means that practitioners will lack critical understanding of the foundations on which they are building.[15]

Moreover, it is not just software engineers who talk about the automobile. How often have we heard one personal computer or another described as the "Model T of computing?" What would it mean to take the claim seriously? What would a personal computer have to achieve to emulate the Model T as a technological achievement and a social and economic force? Would it be useful for historians of computing to take the Model T as a historical precedent? If not, what is a useful historical precedent? Are there perhaps several precedents?

An Agenda for History of Computing

So, maybe the history of computing is not so easy, but what's to be done? Let me conclude by setting out an agenda, for which I claim neither completeness nor objectivity. For one thing, it reflects my own bias toward software rather than hardware.

We need to know more than we do about how the computing industry began. What was the role of the government? How did both older companies and startups identify a potential market and how did the market determine their products? What place did research and development occupy in the companies' organization, and how did companies identify and recruit staffs with the requisite skills?

We need to know about the origins and development of programming as an occupation, a profession, a scientific and technological activity. A proper history here might help in separating reality from wishful thinking about the nature of programming and programmers. In particular, it will be necessary for historians of computing to embed their subject into the larger contexts of the history of technology and the history of the professions as a whole. For example, in *The System of Professions* Andrew Abbott argues that

> it is the history of jurisdictional disputes that is the real, the determining history of the professions. Jurisdictional claims furnish the impetus and the

pattern to organizational developments. . . . Professions develop when jurisdictions become vacant, which may happen because they are newly created
or because an earlier tenant has left them altogether or lost its firm grip on
them.[16]

That is, the professions as a system represent a form of musical chairs
among occupations except that chairs and participants may be added as
well as eliminated. Not all occupations are involved. Competition takes
place among those that on the basis of abstract knowledge "can redefine
[their] problems and tasks, defend them from interlopers, and seize new
problems." Although craft occupations control techniques, they do not
generally seek to extend their aegis.

One only has to page through the various computing journals of the
late 1950s and early 1960s to see conflicts over jurisdiction reflecting
uncertain standing as a profession. Whose machine was it? Soon after the
founding of the ACM, it turned its focus from hardware to software, ced-
ing computing machinery to the IRE. Not long thereafter, representatives
of business computing complained about the ACM's bias toward scientific computing and computer science. Numerical analysts scoffed at "computerologists," inviting them to get back to the business at hand. One
does not have to look hard to find complaints in other quarters that
computing was being taken over by academics ignorant of the problems
and methods of "real" programming. Similar tensions underlay discussions of a succession of ACM committees charged with setting the curriculum for computer science.

Following from that is the need for histories of the main communities
of computing: numerical analysis, data processing, systems programming, computer science, artificial intelligence, graphics, and so on. When
and how did the various specialties emerge, and how did each of them
establish a separate identity as evidenced, say, by recognition as a SIG by
the ACM or the IEEE, or by a distinct place in the computing curriculum?
How has the balance of professional power shifted among these communities, and how has the shift been reflected in the technology?

The software crisis proclaimed at the end of the 1960s is still with us,
despite the accelerated expansion of software engineering in the 1970s and
1980s. The origins and development of the problem and the response to it
provide a rare opportunity to trace the history of a discipline shaping itself.
It did not exist in 1969. Throughout the 1970s and early 1980s, keynote
speakers and introductions to books repeatedly asked, "Are we there yet?,"
without making entirely clear where "there" was. Today, software engineering has a SIG, its own *Transactions*, its own IEEE publication, an
ACM approved curriculum, and a growing presence in undergraduate and

graduate programs. One only has to compare Barry Boehm's famous article of 1973 with reports about DoD software in any issue of *Software Engineering Notes* to doubt that the burgeoning of software engineering reflects its success in meeting the problems of producing reliable software on time, within budget, and to specifications. How, then, has the field grown so markedly in twenty years?[17]

Software takes many forms, and we have begun the history of very few of them, most notably programming languages and artificial intelligence. Still awaiting the historian's attention are operating systems, networks, databases, graphics, and embedded systems, not to mention the wealth and variety of microcomputer programs. This is HOPL II; we still await HOS I (operating systems) or any of a host of HOXs. SIGGRAPH undertook a few years ago to determine the milestones of the field, but there has been little or no work since then. In each of these areas, history will have to look well beyond the software itself to the fields that stimulated its development, supplied the substantive methods, and in turn incorporated computing into their own patterns of thinking.

Finally, there remains the elusive character of the "Computer Revolution," first proclaimed by Edmund C. Berkeley, then editor of *Computers and Automation*, back in 1962 and subsequently heralded by a long line of writers, both in and out of computing. Clearly something epochal has happened. Yet, as I pointed out several years ago in an article, it would be hard to demonstrate for the computer, in whatever form, as pervasive an impact on ordinary people's lives as Robert and Helen Lynd were able to document for the automobile in their famous study of Muncie, Indiana in 1924. But here I am, back at the automobile again. Maybe I've been spending too much time with computer people.[18]

CHAPTER FOUR

The Histories of Computing(s)

T HE "DIGITAL" in the title of the lecture series from which this paper derives points to the future of the humanities, which for the moment remain still largely "analogue." I can't claim strong credentials when it comes to looking toward the future. During my final year at Harvard in 1959–1960, I had a job as a computer programmer for a small electronics firm in Boston. It involved writing code for a Datatron 204, soon to become through acquisition the Burroughs 204, a decimally addressed, magnetic drum machine. Programming it meant understanding how it worked, since it was just you and the computer: no operating system, no programming support. Six or seven months of that persuaded me that computers were not very interesting, nor did they seem to me to have much of a future. So I abandoned my thoughts of going into applied mathematics and became a historian instead. With foresight like that, it was probably a good choice.

I'm not sure my foresight improved when I returned as historian to computing in the early 1980s, at least not enough to make the right investments. Yet, I did have enough critical understanding of what was happening to distrust the utopian promises of the time. Henry Adams showed us in his *Education* how the past can be a good place to look for the future. History helps to us to know where we might be going by establishing where we are and how we got here. The future is largely the result of our present momentum. There are surprises, of course, things we don't see coming. Flash Gordon and Buck Rogers were rocketing among the planets in the 1930s without the aid of computers, which we now know to be indispensable to the enterprise and which made a reality of fiction. But even surprises wind up having a history, first because the new most often comes incrementally embedded in the old and because in the face of the unprecedented we look for precedents. It is the only way we can understand what is new about something, or the way in which it is new, and adjust our thinking to it.[1]

It is especially important to bear that in mind with respect to computers and computing, because they have always been surrounded by hype (it was—and may still be—the only way to sell them), and hype hides history. From the very beginning people in the field have been engaged in instant historical analysis aimed at declaring a new epoch, a radical disjuncture. Edmund C. Berkeley, the author in 1948 of *Giant Brains, or Machines That Think*, first announced the "computer revolution" in a book with that title in 1962. Identifying it as an aspect of the "second industrial revolution," Berkeley combined a gentle introduction to computers with fretting over the social implications of the automation of thought. Since his book, the word "revolution" has appeared regularly in tandem with "computer" or "information," where the latter term has become all but identical with what computers store and manipulate. Were one to take this literature at face value, the job of the historian would seem clear: to chronicle the revolution—or, rather, revolutions, because no area of computing is complete without causing a revolution—and to rank it among the other great revolutions of history: the agricultural revolution, the scientific revolution, the industrial revolution.[2]

Historians prefer to judge revolutions in retrospect, after the dust has settled. Revolutions aren't what they used to be. The events of the past decades have shown us how hard it is to erase or escape a people's history. As an historian of science I'm watching a subject I've taught for more than thirty years, the Scientific Revolution, declared a non-event. More important, perhaps, most declarations of the "computer revolution" have rested on future promises rather than on present or past performance: "See what computers are on the verge of doing. It will be revolutionary!" and "When computers start thinking for themselves, what is going to become of us?" Anyone familiar with the literature of computing over the past forty years knows that the field has been long on promises and short on performance. The literature is filled with revolutionary breakthroughs postponed owing to technical difficulties.[3]

History is the record of our collective experience, our social memory. We turn to it, as we do to our personal experience, consciously when we meet new situations, unconsciously as we live day-to-day. My research into formation of two new subjects, theoretical computer science and software engineering, reveals how people engaged in new enterprises bring their histories to the task, often different histories reflecting their different backgrounds and training. The creators and early practitioners of these fields all came from somewhere else. In a recent article, "Finding a History for Software Engineering," I have tried to show how efforts to define and articulate a new engineering discipline for software have rested on practitioners' understanding of the history of other fields of engineering. It is incorporated in our institutions, most particularly our schools, where it is taught consciously in history courses, unconsciously in every other part of the curriculum. It is embodied in the artifacts we use, in the customs we follow, in the language we speak. Whether we want to or not, we use history. As in the case of personal memory, the question is whether we use it well. That is a matter both of getting the history right and of getting the right history. As Rob Kling and Walt Scacchi argue in their studies of the social patterns of computing and their impact on systems design, "history of commitments constrains choices"—even, or especially, revolutionary choices.[4]

Decentering the Machine

With some recent exceptions, the history of computing has been centered on the machine, tracing its origins back to the abacus and the first mechanical calculators and then following its evolution through the generations of mainframe, mini, and micro. Alongside this machinic main thread run accounts of scientific calculations; statistics and tabulations; and the growing informational needs of business, industry, and government, all converging in the mid-1940s on the electronic digital computer and then spreading out again in new forms shaped by it, in the end to be tied together by the internet (fig. 1). Once invented, the computer evolves naturally into the PC as its present most visible form, rather than into a variety of coexisting, mutually supportive forms (as if mainframes disappeared with the invention of the minicomputer). Its progress is inevitable and unstoppable, its effects revolutionary.

Chronicling the revolution, that machine-centered history reinforces the hype and with it what one might call the "impact theory" of the relation of technology and society. There is society strolling along, minding its own business, and, wham!, it gets impacted and is left reeling by a revolutionary technology, which changes everything overnight or in some similarly short time. "The ominous rumble you sense is the future coming at us," wrote one management systems expert in 1953. From that perspective, society breaks up into two classes: those who are on the train and those who are not, and the latter are hopeless (or, as one enthusiast recently put it with a curious nod to Engels, consigned to the dustbin of history). In our field, you're either a digeratus or a dinosaur.[5]

But history of that sort should give us pause on at least two counts. First, computing is a technology (or a constellation of technologies) and, however revolutionary, should have the same sort of richly contextual

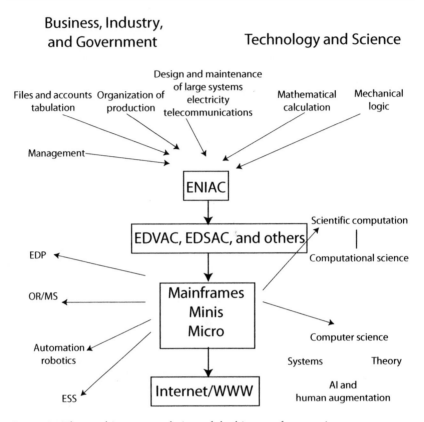

Figure 1   The machine-centered view of the history of computing.

history that other revolutionary technologies have—such as the steam-powered factory and the automobile. The question is not whether new technology involves social change, but how it does. In particular, it is a question of agency. As a form of technological determinism, the impact theory leaves people reacting to technology, rather than actively shaping it. Much of the thoughtful history of technology over the past twenty years has aimed at getting people back into the picture or, to change the metaphor, into the driver's seat. The devices and systems of technology are not natural phenomena but the products of human design, that is, they are the result of matching available means to desired ends at acceptable cost. The available means ultimately do rest on natural laws, which define the possibilities and limits of the technology. But desired ends and acceptable costs are matters of society. Given a set of possibilities, what do we want to do, and what are we willing to pay (in money, time, effort, and trade-offs) to do it?

Second, whereas other technologies may be said to have a nature of their own and thus to exercise some agency in their design, the computer has no such nature. Or, rather, its nature is protean; the computer is—or certainly was at the beginning—what we make of it (or now have made of it) through the tasks we set for it and the programs we write for it. Let me take a moment to explain.

The Protean Machine

The new appears to us in contexts defined by the old, by history, or rather histories. Look at a picture of the ENIAC, the first electronic digital calculator (not yet a full computer, but its immediate predecessor). It is a new device constructed from existing components, as the scheme by Arthur Burks shows (fig. 2). Those components embody its history, or rather the history of which it was a product. At this point it was merely doing electronically, albeit for that reason much more quickly, what other devices of the period were doing electromechanically and previous devices had done mechanically, and one can find indications of that throughout its design.

Only in the next iteration of its design, the EDVAC, could it do things no earlier machine had been able to do, namely, make logical decisions based on the calculations it was carrying out and modify its own instructions. The elements of that design have another history, as different from that of ENIAC as the schematic version (fig. 3) is from its circuit diagrams. That is the history of logic machines, reaching back to Leibniz and running through Boole and Turing. The combination of those two histories made the computer in concept a universal Turing machine, limited in practice by its finite speed and capacity. But making it universal, or general-purpose, also made it indeterminate. Capable of calculating any logical function, it could become anything but was in itself nothing (well, as designed, it could always do arithmetic).[6]

Communities of Computing

In the early years, then, as the computer moved out of the science and engineering laboratory, it did not bring much history of its own with it. It could not dictate how it was to be used and hence could not have an "impact" until people figured out what to do with it. Mechanical calculation had a history; symbolic processing did not. Thus the history of computing is the history of what people wanted computers to do and how people designed computers to do it. That may not be one history, or at least it may not be useful to treat it as one. Different groups of people

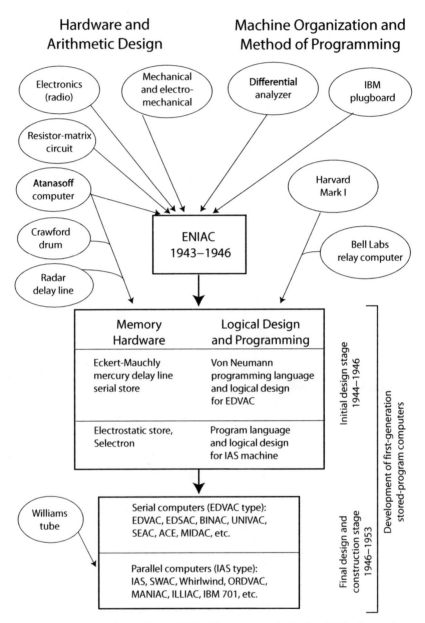

Figure 2 Sources of ENIAC and EDVAC, as set out in Arthur W. Burks and Alice R. Burks, "The ENIAC: First General Purpose Electronic Computer," *Annals of the History of Computing* 3, no. 4 (October 1981): 310–389.

John von Neumann et al., EDVAC Architecture

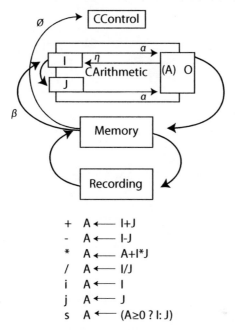

$$+ \quad A \longleftarrow I+J$$
$$- \quad A \longleftarrow I-J$$
$$* \quad A \longleftarrow A+I*J$$
$$/ \quad A \longleftarrow I/J$$
$$i \quad A \longleftarrow I$$
$$j \quad A \longleftarrow J$$
$$s \quad A \longleftarrow (A \geq 0 \, ? \, I: J)$$

Figure 3   Architecture of EDVAC (schematic diagram based on J. von Neumann, "First draft of a report on the EDVAC").

saw different possibilities in computing, and they had different experiences as they sought to realize those possibilities. One may speak of them as "communities of computing," or perhaps as communities of practice that took up the computer, adapting to it while they adapted it to their purposes. The chart shows the major groups (fig. 4).

The first, of course, is the scientists and engineers for whose needs the computer had been created. Their practice guided the earliest designs of computers, the series of one-off machines built at government and university sites. They are the community for which the first high-level programming language, FORTRAN, was created, a language which continues in use down to the present.

The second group comprises the field of data processing, the first commercial extension of the computer. Recent work by Martin Campbell-Kelly, James Cortada, and others (themselves working from an older literature) reveals a rich history going back more than a century. That history explains why IBM took time to decide to go into the computer business and why the business took the form it did once the decision had been made.

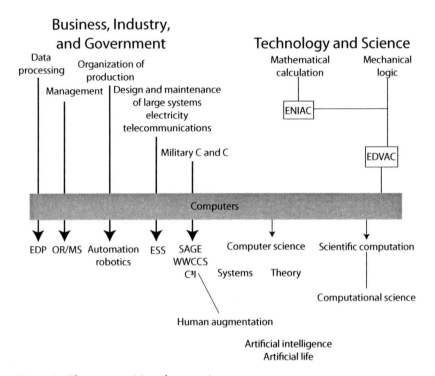

Figure 4   The communities of computing.

Tightly connected with data processing is the field of management science, dating back to the work of Frederick W. Taylor and others at the turn of the last century, and beyond that perhaps to Charles Babbage's *On the Economy of Machinery and Manufactures*. Operations research (or in England, operational research) brought a new mathematical dimension to the field as practitioners sought to apply methods of tactical evaluations to business and industry.

Computers also fit into the agenda of industrial engineering in the area of automation and control of production flow, which was given its twentieth-century shape by Henry Ford but which can arguably be traced back to Matthew Boulton and James Watt. Indeed, as just noted, it is here, rather than to the Analytical Engine, that one should be looking to place Charles Babbage into the genealogy of computing.[7]

Having already created some of the first electromechanical computing devices to address problems of network analysis and switching systems, the communications industry quickly adopted the electronic computer and began the process of building it into our environment.

Automation of military command and control systems (SAGE and WW-MCCS) reflects the history of modern warfare, which reaches back several centuries. In light of recent literature on the subject, we need a fuller and more subtle understanding than we have of the history of the community of practitioners in the military, to compensate for the visions of the future designed for it by researchers outside it. An important spin-off of this line of research was the field of human-computer interaction, aimed at the augmentation of human skills by the computer. As David Mindell has shown recently in *Between Human and Machine*, that field too had a history of its own before the computer.[8]

New fields emerged alongside those that predated the computer. Although there was at first some disagreement about whether computers were sufficiently different from early calculating machines to constitute a subject of study in themselves, the work of John von Neumann and Alan Turing pointed to the theoretical potential of the device. Moreover, the need to develop systems to both make the computer easier to use and to keep it running efficiently also opened a range of theoretical and technical issues that prompted the emergence of the new subject of computer science or informatics, as it is known in the non-anglophone world.

Until recently the history of computing in these fields has been written in terms of the machine and its impact (revolutionary, of course) on them. The emphasis has lain on what the computer could do rather on how the computer was made to do it. As I said, in many cases their existence as fields of computing has led to their being written into the history leading up to the computer as if the purposes to which they put the computer had somehow previously constituted a demand for which the computer had been created. Within the last few years, however, that view has begun to shift. In *Information Technology as Business History*, James Cortada justifies a chapter on the "100-year history of mechanical devices used in modern times to do data processing" by admonishing:

> Why do we care? Many of the ways you and I receive computers, buy them, and use this technology were worked out in companies that existed ten, thirty, even fifty years before the first commercial computer was available. Did you know that National Cash Register (NCR) and Burroughs—in time both peddlers of computers—had been selling information-handling machines over 100 years ago—or, that International Business Machines Corporation (IBM) has been around since before World War I? Historical context is so important; computers have a lot of it, and its patterns of behavior are described in this chapter.[9]

Tom Haigh, whose work on the history of information systems is beginning to appear, reinforces the point:

Work by historians such as Martin Campbell-Kelly, JoAnne Yates, and William Aspray has consistently shown that the computer industry was, more than anything else, a continuation of the pre-1945 office equipment industry—and in particular of the punched card machine industry. Their careful exploration of computer technology and the dynamics of the computer hardware industry leave little doubt that IBM's eventual dominance of the computer industry owes as much to the events of the 1930s as to those of the 1960s. This is in itself a major departure from the perception, common during the 1950s and common today, that each new generation of computer equipment is a revolutionary technology without historical roots, a breakthrough plucked fully formed from the forehead of (to mix a metaphor) Prometheus.[10]

When looking at the introduction of computing into the business world, Haigh insists that we must break away from a focus on computing.

The use of computer technology in a particular social space (such as the laboratory, office, or factory) cannot be addressed without also studying the earlier history of this setting, the people in it, and the objectives to which the machine is put. So, while coherent one-volume histories of the computer hardware industry and its technologies can be written, it seems unlikely that we can produce a single coherent narrative about the use of computers or of associated tasks such as analysis, programming, or operation.[11]

Approaching data processing from this wider perspective reveals a history quite different from the current history of computing.

Taking a similarly inspired approach in *The Government Machine*, Jon Agar gives an idea of how different that history looks for the development of computing in Britain. Agar looks back to the eighteenth century, when political thinkers began to speak of "the machinery of government," and follows the development of that metaphor, as it led to the ever-widening collection of data about the population and the introduction of machines of various sorts to record and manipulate the data. Skeptical of the notion of a computer-based "information revolution" and the historical discontinuity it implies, Agar focuses rather on "the humans who promoted machines" and on the technocratic "vision of government" that conditioned the adoption of office technologies, including the computer. Indeed, he maintains, "the Civil Service, as a general-purpose universal machine, framed the language of what a computer was and could do."[12]

As Cortada, Haigh, and Agar suggest, the histories and continuing experience of the various communities show that they wanted and expected different things from the computer. They encountered different problems and levels of difficulty in fitting their practice to it. As a result, they created different computers or (if we may make the singular plural) computings. To do so, they had to determine which aspects of their practice were suitable for automation, they had to build computational models of those aspects, and they had to write the programs that implemented those models. None of this was straightforward, except where it was trivial. From the case studies we do have, we can guess that it was a matter of negotiation, both among practitioners about the nature of their practice and between practitioners and the realities of the current technology, often in the shape of non-practitioner technicians.

Until recently, histories of computing have largely ignored that process of negotiation, and that should concern more than historians. From the early 1950s down to the present, various communities of computing have translated large portions of our world—our experience of it and our interaction with it—into computational models to be enacted on computers, not only the computers that we encounter directly but also the computers that we have embedded in the objects around us to make them "intelligent" or even, as Yorick Wilks would have it, "companionable." They have increasingly made computers the medium of our working (in the broadest sense of the term) in the world. In doing so, they have (re)shaped not only their own practice, but also computers and their adaptation by others. So far, we know little of the process by which they have done it. We have the story of where the physical devices came from, how they have taken their current form, and what differences they have made. But we remain largely ignorant about the origins and development of the dynamic processes running on those devices, the processes that determine what we do with computers and how we think about what we do. The histories of computing will involve many aspects, but primarily they will be histories of software.

Worlds of Software

Historians of computing have only begun to tackle the history of software. We've stuck pretty close to the machine. We know a great deal about the history of programming languages and considerably less about the history of operating systems, databases, and other varieties of systems software. We have scarcely scratched the surface of applications software, the software that actually gets things done outside the world of the computer itself.

A recent conference at the Nixdorf Museum in Paderborn, Germany, attempted to map the history of software, considering it as science, engineering, labor process, reliable artifact, and industry, with a look at the question of how one exhibits it in a museum. The focus lay on software and its production as a general phenomenon. What the conference missed was software as model, software as experience, software as medium of thought and action, software as environment within which people work and live. It did not consider the question of how we have put the world into computers.[13]

That process has not been easy or straightforward. If it appears so, it is because so far we have concentrated on the success stories and told them in a way that masks the compromises between what was intended and what could be realized. Programming is where aspiration meets reality. The enduring experience of the communities of computing has been the huge gap between what we can imagine computers doing and what we can actually make them do. There have been (and continue to be) some massive failures in software development, which have cost money, time, property, and even lives. Indeed, since the late 1960s people in the field have spoken of a "software crisis." Yet, except for Frederick P. Brooks's famous *Mythical Man-Month*, we do not have substantive accounts of those failures, not even in the software engineering classroom, where we might expect to find them. As software engineer/historian James Tomayko points out, other branches of engineering learn from their mistakes. As Henry Petroski puts it in the title of one of his books on the role of failure in design, "to engineer is human." Software will look more human when we take seriously the difficulties of designing and building it.[14]

Let me say a bit more about the "software crisis." It emerged in the late 1960s as an increasing number of large projects experienced large cost overruns, missed deadlines, and failures to meet specifications. Some had to be canceled altogether. One response was calls for a discipline of "software engineering" that would, in the words of a deliberately provocative definition, base "software manufacture . . . on the types of theoretical foundations and practical disciplines that are traditional in the established branches of engineering." (That is, of course, a historical program.) At first, practitioners sought a solution in the computer and looked to the improvement of their tools and at more effective project management. A great deal of effort went into developing high-level programming languages and diagnostic compilers for them. The languages were specifically designed to foster good programming practices (read, proper ways to think about programming). Similarly, practitioners sought to bring the experience of industrial engineering to bear on software production, with an eye toward automating it in the form of a "software factory," a programming environment that would leave the programmer little choice but to do it right.[15]

The tools clearly got better. Once a project gets down to the actual programming, things go relatively smoothly. But, as Brooks has since pointed out, that is not where the real problems have lain, or rather problems at that level were only "accidental." Almost from the start studies showed that the bulk of the errors occurred at the beginning of projects, before programming ever began (or should have begun). The errors were rooted in failures to understand what was required, to specify completely and consistently how the system was supposed to behave, to anticipate what could go wrong and how to respond, and so on. As many as two-thirds of the errors uncovered during testing could be traced back to inadequate *design*; the longer they remained undetected, the more costly and difficult they were to correct.[16]

Design is not primarily about computing as commonly understood, that is, about computers and programming. It is about modeling the world in the computer, about computational modeling, about translating a portion of the world into terms a computer can "understand." Here it may help to go back to the protean scheme to recall what computers do. They take sequences, or strings, of symbols and transform them into other strings. The symbols and the strings may have several levels of structure, from bits to bytes to groups of bytes to groups of groups of bytes, and one may think of the transformations as acting on particular levels. But in the end, computation is about rewriting strings of symbols.

The transformations themselves are strictly syntactical, or structural. They may have a semantics in the sense that certain symbols or sequences of symbols are transformed in certain ways, but even that semantics is syntactically defined. Any meaning the symbols may have is acquired and expressed at the interface between a computation and the world in which it is embedded. The symbols and their combinations express representations of the world, which have meaning to us, not to the computer. It is a matter of representations in and representations out. What characterizes the representations is that they are operative (cf. algebra as operative symbolism). We can manipulate them, and they in turn can trigger actions in the world. What we can make computers do depends on how we can represent in the symbols of computation portions of the world of interest to us and how we can translate the resulting transformed representation into desired actions. We represent in a variety of forms a Boeing 777: its shape, structure, flight dynamics, controls. Our representations not only direct the design of the aircraft and the machining and assembly of its components, but they then interactively direct the control surfaces of the aircraft in flight. That is what I mean by "operative representation."

So putting a portion of the world into the computer means designing an operative representation of it that captures what we take to be its essential features. That has proved, as I say, no easy task; on the contrary it has proved difficult, frustrating, and in some cases disastrous. It has most recently moved to a high-priority problem at the U.S. National Science Foundation, which has just announced a program aimed at exploring the "science of design." Where that will go is anyone's guess at the moment, and I'm a poor prognosticator. What is clear is that historians of computing have inherited the problems to which it is addressed. If we want critical understanding of how various communities of computing have put their portion of the world into software, we must uncover the operative representations they have designed and constructed, and that may prove almost as difficult a task.[17]

Reading Machines, Real and Virtual

What makes it difficult is precisely that the representations are operative. Ultimately it is their behavior rather than their structure (or the fit between structure and behavior) that interests us. We do not interact with computers by reading programs; we interact with programs running on computers. The primary source for the historian of software is the dynamic process, and, where it is still available, it requires special techniques of analysis. Programs and processes are artifacts, and we must learn to read them as such.

A brief digression is necessary here. My return to computing occurred when I started teaching the history of technology. Up to that point I had taught history of science from antiquity through the scientific revolution, and I based my courses almost entirely on primary sources. My students read Plato, Aristotle, Aquinas, Copernicus, Galileo, and so on. So I looked for primary sources on mills, steam engines, automobiles, and computers, and had great difficulty in finding what I wanted. It soon dawned on me that I was looking for the wrong thing in the wrong place. The sources I needed were not texts about these machines, but the machines themselves, to be found not in a library but in a museum. Technology is not a literate enterprise, but a visual, tactile one. Its practitioners think not with words but, as Derek de Solla Price so deftly phrased it, "with their fingertips." Henry Ford put it another way:

> There is an immense amount to be learned simply by tinkering with things. It is not possible to learn from books how everything is made—and a real mechanic ought to know how nearly everything is made. Machines are to a mechanic what books are to a writer. He gets ideas from them, and if he has any brains he will apply those ideas.[18]

What holds for the practitioners is true also of those who use technologies, or as Langdon Winner insists, *live* them. We do not read about them;

we act with, in, and through them. Conversely, their design assumes that we know certain things or can learn them. They are artifacts of our *culture*, embodying both its explicit and its tacit knowledge.[19]

In the realm of computing, the notion of reading artifacts transfers more or less readily to the machine itself. In *The Soul of a New Machine*, Tracy Kidder relates the story of computer designer Tom West sneaking a peek at a competitor's design [and seeing its organizational chart reflected inside].* But reading the software is even trickier, because we can't "pull the boards." We must learn to interrogate the artifact in action, and here we need help from sociologists, anthropologists, and the HCI community, whose studies of current users may suggest interpretive approaches to the past.

Here historians of software face a problem caused by the rapid rate of obsolescence that has characterized computing from the outset. For the early period that particularly interests us, we cannot read the dynamic artifacts, because we no longer have the platforms, the machines and operating systems, on which the software ran. In some cases the disappearance of the platform has meant the loss of the software as well. Given present trends, there is no reason to think that will not be the case with more recent software unless we embark on a systematic program to archive software and hardware in ways that allow retrieval of the dynamic artifact. It is a bit ironic that in an age that seems overwhelmed by information, the history of the technology designed to manage it will itself be hampered by lack of information, as a crucial body of primary sources is lost to us through the disappearance of the machines on which they were enacted.[20]

In the absence of the running process, the next best thing is the program text, the source code. Again, as historians we inherit the unsolved problems of the subject we are studying. As Christopher Langton, lead proponent of artificial life, has put it:

> We need to separate the notion of a formal specification of a machine—that is, a specification of the *logical* structure of the machine—from the notion of a formal specification of a machine's behavior—that is, a specification of the sequence of transitions that the machine will undergo. In general, we cannot derive behaviours from structure, nor can we derive structure from behaviours.[21]

Despite impressive work in the mathematical theory of program syntax and semantics, we have no means of deriving the dynamic process from the static program, in the sense of being able to determine from the latter the

---

* Editorial note: Mahoney included here the lengthy quotation from Kidder also found in "Issues in the History of Computing." It has been removed to avoid duplication.

state of the former at some particular point in the computation. Perhaps that should not surprise us. If we did, we would not need the computer.

Nonetheless, if we have the program, we can try to reconstruct from it what the process looked like, and we can learn to analyze the text to discover the structures and operations of the computational model and, through them, to gain some sense of the understandings and intentions of its designers. Alan Kay, the creator of Smalltalk, offered an example through his own discovery of how the language Simula worked.* Kay's experience suggests what faces the historian, and there is again some irony in it. It has been the common lament of management that programs are built by tinkering and that little of their design gets captured in written form, at least in a written form that would make it possible to determine how they work or why they work as they do rather than in other readily imaginable ways. Moreover, what programs do and what the documentation says they do are not always the same thing. Here, in a very real sense, the historian inherits the problems of software maintenance: the farther the program lies from its creators, the more difficult it is to discern its architecture and the design decisions that inform it.

But at least we have the program texts. Or do we? It is not clear, in part perhaps because for many of the communities of computing outside computer science we have not yet begun to look. When government and industry went looking in the late 1990s in anticipation of Y2K, the results were dismaying. But that is another, different problem.

History, Computing, and the Humanities

What does this mean for digital scholarship and our digital future? Let me go back to the beginning. Look at a picture of the EDSAC at Cambridge, the first operational computer, and then look at a laptop. A room full of electronic equipment has become a typewriter connected to a TV screen by a black box, with sockets in the back for attachments. What separates the two devices is not evolution but social construction, a lot of it. The difference is not the result so much of working out principles as of pursuing the possibilities of practice.

As striking as is the contrast in external form of the two computers, what really separates the EDSAC from my laptop is something on the order of 3 gigabytes of stored programs, a large number of which simply establish a platform for the programs that I use to do what I want. For the

---

* Editorial note: Mahoney included here the lengthy quotation from Kay also found in "Issues in the History of Computing." It has been removed to avoid duplication.

lecture on which this paper is based, I ran a single application to project images to the screen. Enabling that application were some two dozen processes, each in itself a program of some complexity. All these programs reflect the histories of the communities from which they come. The operating system, for example, embodies the history of corporate organizations designed to distribute responsibilities and authority in a hierarchical structure. The graphical user interface, known for its main features as "WIMP" (windows, icons, mouse, and pull-down menus), emerged from the human augmentation community, with its roots in behaviorist perceptual psychology and military command and control systems. Microsoft PowerPoint reflects the adaptation of computer-assisted design to the needs of management systems and the corporate boardroom. The communications community provides the networking. And so on. To "use" a "personal" computer today is, despite its much hyped origins in the counterculture, to work in a variety of environments created by a host of anonymous people who have made decisions about the tasks to be done and the ways they should be done. As most of us use a computer, it is no more personal than a restaurant: you can have anything you want on the menu, cooked the way the kitchen has prepared it.

Is that bad? No, it is the nature of a technological system. It is the price of computing power. I've worked on a computer without an operating system or a library of programs. The experience drove me from the field. I'm as happy to use the current computing technology as I am to fly on a 777, drive a maintenance-free automobile, or wear no-iron shirts. It is not bad, but it does have implications for the critical, reflective use of computers, especially in the humanities. It means that the computer as tool and medium is not neutral, but rather informs (or, as Bolter and Grusin put it, re-mediates) the work that one does with it, if only by setting possibilities and limits on what can be done (or even thought). It calls for critical awareness. Like the historians of computing, digital scholars must learn to read software to elicit the history and practice that it embodies.

We must be aware because, by and large, we in the humanities have had to borrow our computing from other communities. In the beginning, we simply couldn't afford it on our own. Since then, we have not formed a significant community giving shape to computers by creating our own computing—except perhaps most recently in the design of mark-up languages and the semantic web. My experience at Princeton is arguably typical. Until the mid-1980s, the scientists and engineers owned the mainframe and the minicomputers, allowing the humanists a bit of time on the side. The personal computer came onto the Princeton campus in the mid-1980s as part of an IBM-sponsored project to make the PC an educational resource. IBM supplied machines and basic software; the universities were supposed to generate applications for teaching and research. But, unless the project aimed at a single target, such as a graphically displayed database of U.S. census data down to the county level, not much happened. One reason was that we were trying to develop educational applications using software designed for business. The same was true in the primary and secondary schools, indeed to a greater degree, since they lacked any resources for independent development. (I happened to be a member of our local school board at the time, so could observe developments at both levels.) Either one used the educational software packages of the day (the less said about them, the better), or one tried to adapt to the classroom software originally developed for the business office. Tools embody history. We were trying to work with other people's histories, unaware of what that meant.

Our own history poses a difficult challenge to the computing community. At a workshop for a large project, the National Initiative to Network the Cultural Heritage (NINCH), jointly sponsored by the American Council of Learned Societies and the National Academy of Engineering, humanists and computer scientists gathered to talk about what humanists do, how they do it, what they'd like to be able to do and how they'd like to be able to do it, and how those wishes might constitute research projects for the computer scientists. I was moderating the discussion among the historians and sensed an uncharacteristic humility among my colleagues, who seemed worried about finding something difficult enough to get the attention of the computer scientists. I tried to reassure them that should not be a concern. Our problems are far too difficult for the computer scientists. We need to find something simple enough for them, yet interesting enough for us. Our main tasks involve the sophisticated use of a natural language, or even of several natural languages; the discerning of subtle patterns of shape, color, texture, and sound. Ours is an enterprise of metaphor, analogy, allusion, ambiguity, etc. These are not things that have so far lent themselves to computational modeling. The humanities involve those aspects of human thinking and cognition that have so far confounded artificial intelligence.

The future of digital scholarship depends on whether we can now design computational models of the aspects of the world that most interest us. That, in turn, calls for reflection on our current practice. In seeking to do things in new ways with a computer, it is useful to clarify how we do them now and how we came to do them that way and not otherwise. Indeed, as Willard McCarty has pointed out, grappling with new technology reminds us that we have been using technology all along: pencil, pen, and paper are technologies, and all forms of calculation involve instruments. In *Remediation*, a study of the digital medium in the arts and literature, Jay Bolter and Richard Grusin play on the ambiguity of the title, arguing that every re-mediation, every transfer of an activity from one medium to another, rests on a claim of remediation, of making things better. What are we seeking to remedy in the remediation of scholarship in the humanities? What do we want to do that, to borrow from Andrew Marvell, we have lacked "world enough and time" to do? The computer, a coy mistress indeed, offers to supply that world and time, provided that we know how to do what we want and can explain it in terms the computer can understand. Humans have to think hard about both questions. Compared to them, the programming will be easy.[22]

PART TWO

Constructing a History for Software

CHAPTER FIVE

Software: The Self-Programming Machine

I N MAY 1973 *Datamation* published a Rand report filed six months earlier by Barry Boehm and based on studies undertaken by the Air Force Systems Command, which was concerned about the growing mismatch between its needs and its resources in the design and development of computer-based systems. Titled "Software and Its Impact: A Quantitative Assessment," the article attached numbers to the generally shared sense of malaise in the industry: software was getting more and more costly. Drawing on various empirical studies of programming and programmers undertaken in the late 1960s, Boehm tried to indicate where to look for relief by disaggregating the costs into the major stages of software projects. Perhaps the most striking visualization of the problem was a graph with a flattened logistic curve illustrating the inversion of the relative costs of hardware and software over the thirty-year period 1955–1985. Whereas software had constituted less than 20 percent of the cost of a system in 1955, current trends suggested that it would make up over 90 percent by 1985. At the time of Boehm's study, software's share already stood at 75 percent.

Boehm's article belongs to the larger issue of the "software crisis" and the origins of software engineering, to which I shall return presently, but for the moment it also serves to make a historiographical point. Software development has remained a labor-intensive activity, an art rather than a science. Indeed, that is what computer people have found so troublesome and some have tried to remedy. Boehm's figures show that by 1970 some three-quarters of the productive energies of the computer industry were going into software. By then at the latest, the history of computing had become the history of software.

At present the literature of the history of computing does not reflect that fact. Except perhaps for the major programming languages, the story of software has been largely neglected. The history of areas such as operating systems, databases, graphics, real-time and interactive computing still lies in past survey articles, prefaces of textbooks, and retrospectives by the people involved. When one turns from systems software to applications programming, the gap widens. Applications, after all, are what make the computer worth having; without them a computer is of no more utility or value than a television set without broadcasting. James Cortada has provided a start toward a history of applications through his quite useful bibliographic guide, but there are only a few studies of only the largest and most famous programs (SAGE, SABRE, ERMA, etc.). We have practically no historical accounts of how, starting in the early 1950s, government, business, and industry put their operations on the computer. Aside from a few studies with a primarily sociological focus in the 1970s, programming as a new technical activity and programmers as a new labor force have received no historical attention. Except for very recent studies of the origins and development of the Internet, we have no substantial histories of the word processor, the spreadsheet, communications, or the other software on which the personal computer industry and some of the nation's largest personal fortunes rest.

Software, then, presents a huge territory awaiting historical exploration, with only a few guideposts by which to maintain one's bearings. One guiding principle in particular seems clear: if application software is about getting the computer to do something useful in the world, systems software is about getting the computer to do the applications programming. It is the latter theme that I shall mainly pursue here. Eventually, I shall come back to applications programming by way of software engineering, but only insofar as it touches on the main theme.

Programming Computers

Basically, programming is a simple, logical procedure, but as the problems to be solved grow, the labor of programming also increases, and the aid of the computer is enlisted to devise its own programs.[1]

 —Werner Bucholz

The idea of programs that write programs is inherent in the concept of the universal Turing machine, set forth by Alan M. Turing in 1936. He showed that any computation can be described in terms of a machine shifting among a finite number of states in response to a sequence of symbols read and written one at a time on a potentially infinite tape. Since the description of the machine can itself be expressed as a sequence of symbols, Turing went on to describe a universal machine which can read the description of any specific machine and then carry out the computation it describes. The computation in question can very well be a description of a universal Turing machine, a notion which John von Neumann pursued to its logical conclusion in his work on self-replicating automata. As a form of Turing machine, the stored-program computer is in principle a self-programming device, limited in practice by finite memory. That limitation seemed overwhelming at first, but in the mid-1950s, the concept of computer-assisted programming began to meet with striking success in the form of programming languages, programming and operating systems, and databases and report generators.

Indeed, that success emboldened people to think about programming languages and programming environments that would obviate the need for programmers in the long run and in the meantime bring them under increasing effective managerial control. By 1961 Herbert A. Simon was not alone in predicting that:

> we can dismiss the notion that computer programmers will become a powerful elite in the automated corporation. It is far more likely that the programming occupation will become extinct (through the further development of self-programming techniques) than that it will become all powerful. More and more, computers will program themselves; and direction will be given to computers through the mediation of compiling systems that will be completely neutral so far as the content of the decision rules is concerned.[2]

Simon was talking about 1985, yet, as we near the millennium, programmers are neither extinct nor even an endangered species. Indeed, old COBOL programmers found renewed life in patching the Y2K problem.

Coincidentally, Simon's remarks were reprinted by John Diebold in 1973, which is just about the point of transition between the successful and the less successful phases of the project of the self-programming computer. By the early 1970s, the basic elements of current systems software were in place, and development efforts since then have been aimed largely at their refinement and extension. With few exceptions, the programming languages covered in the two ACM History of Programming Languages conferences in 1978 and 1993 were conceived before 1975. They include the major languages currently in use for applications and systems programming. In particular, C and UNIX both date from the turn of the 1970s, as do IBM's current operating systems. The graphical user interfaces (GUIs) of Windows and MacOS rest on foundations laid at Stanford in the 1960s and Xerox PARC in the early and mid-1970s. The seminal innovations in both local- and wide-area networking also date from that time. Developments since then have built on those foundations.

By 1973, too, "software engineering" was under way as a conscious effort to resolve the problems of producing software on time, within budget, and to specifications. Among the concepts driving that effort, conceived of as a form of industrial engineering, is the "software factory," either on the Taylorist model of "the one best way" of programming enforced by the programming environment or on the Ford model of the assembly line, where automated programming removes the need for enforcement by severely reducing the role of human judgment. At a conference in August 1996 on the history of software engineering, leading figures of the field agreed only that after almost thirty years, whatever form it might eventually take as an engineering discipline, it wasn't one yet. While software development environments have automated some tasks, programming "in the large" remains a labor-intensive form of craft production.

So we can perhaps usefully break systems software up into programming tools and programming environments on the one hand and software development (or, if you prefer, software engineering) on the other. Both fall under the general theme of getting the computer to do the programming. Both have become prerequisites to getting the computer to do something useful.

Programming Tools

It is a commonplace that a computer can do anything for which precise and unambiguous instructions can be given. The difficulties of programming computers seem to have caught their creators by surprise. Werner Buchholz's optimism is counterbalanced by Maurice Wilkes' realization that he would be spending much of his life debugging programs. On a larger scale, companies that introduced computers into their operations faced the problem of communication between the people who knew how the organization worked and those who knew how the computer worked. IBM had built its electrical accounting machinery (EAM) business in large part by providing that mediation through its sales staff, whose job it was to match IBM's equipment to the customer's business. At first it seemed that computers meant little more than changing the "E" in "EAM" from "Electrical" to "Electronic," but experience soon showed otherwise. Programming the computer proved to be difficult, time-consuming, and error-prone. Even when completed, programs required maintenance in the form of addition of functions not initially specified, adjustment of unanticipated outcomes, and correction of previously undetected mistakes. With each change of computer to a larger or newer model came the need to repeat the programming process from the start, since the old code would not run on the new machine. The situation placed a strain on both the customer and IBM, and together with other manufacturers they therefore shared an interest in means of easing and speeding the task of programming and of making programs compatible with a variety of computers.[3]

In addition to having to work within the confines of the machine's instruction set and hardware protocols, one had to do one's own clerical tasks of assigning variables to memory and of keeping track of the numerical order of the instructions. The last became a systematic problem on Cambridge's Electronic Delay Storage Automatic Calculator (EDSAC) as the notion of a library of subroutines took hold, necessitating the incorporation of the modules at various points in a program. Symbolic assemblers began to appear in the early 1950s, enabling programmers to number instructions provisionally for easy insertion, deletion, and reference and, more important, turning over to the assembler the allocation of memory for symbolically denoted variables. Although symbolic assemblers took over the clerical tasks, they remained tied to the basic instruction set, albeit mnemonic rather than numeric. During the late 1950s macro assemblers enabled programmers to group sequences of instructions as functions and procedures (with parameters) in forms closer to their own way of thinking and thus to extend the instruction set.[4]

The first high-level programming languages, perhaps most famously FORTRAN in 1957, followed over the next three years by LISP, COBOL, and ALGOL, took a quite different approach to programming by differentiating between the language in which humans think about problems and the language by which the machine is addressed. To clerical tasks of the assembler, compilers and interpreters added the functions of parsing the syntax and construing the semantics of the human-oriented programming language and then translating them into the appropriate sequences of assembler or machine instructions. At first, as with FORTRAN, developers of compilers strove for little more than a program that would fit into the target machine and that would produce reasonably efficient code. Once they had established the practicality of compilers, however, they shifted their goals.

In translating human-oriented languages into machine code, compilers separated programming from the machines on which the programs ran: "ALGOL 60 is the name of a notation for expressing computational processes, irrespective of any particular uses or computer implementations," said one of its creators. Subsequently, the design of programming languages increasingly focused on the forms of computational reasoning best suited to various domains of application, while the design of compilers attended to the issues of accurate translation across a range of machines. With that shift of focus at the turn of the 1960s, the development of programming languages and their compilers converged with research in theoretical computer science, first to establish the general principles underlying lexical analysis and the parsing of formal languages, then to implement those principles in general programs for moving from a formal specification of the vocabulary and grammar of a language to the corresponding lexical analyzer and parser, which not only resolved the source program into its constituents and verified its syntactical correctness but also allowed the incorporation of preset blocks of machine code associated with those constituents to produce the compiler itself. By means of such tools, for example, lex and yacc in the UNIX system, a compiler that in the late 1950s would have required several staff-years became feasible for a pair of undergraduates in a semester. By contrast, automatic generation of code, or the translation of the abstract terms of the programming language into the concrete instruction set of the target machine, proved more resistant to theoretical understanding (formal semantics) and thus to automation, especially in a form that assured semantic invariance across platforms.[5]

Systems Software

"Problem-oriented languages," as they were called, were designed to facilitate the work of programmers by freeing them from the operational details of the computer or computers on which their programs would run. The more abstract the language, the more it depended on a programming system to supply those details whether through a library of standard routines or through compilers, linkers, and loaders that fitted the program to the mode of operation of the particular computer. Thus software aimed at shielding the programmer from the machine intersected with software, namely operating systems, meant to shield the machine from the programmer.

Operating systems emerged in the mid-1950s, largely out of concerns to enhance the efficiency of computer operations by minimizing nonproductive time between runs. Rather than allowing programmers to set up and run their jobs one by one, the systems enabled operators to load a batch of programs with accompanying instructions for setup and turn them over to a supervisory program to run them and to alert the operators when their intervention was required. With improvements in hardware, the systems expanded to include transfer and allocation of tasks among several processors (multiprocessing), in particular separating slower input/output (I/O) operations from the main computation. At the turn of the 1960s, with the development of techniques for handling communications between processors, multiprogramming systems began running several programs in common memory, switching control back and forth among them according to increasingly sophisticated scheduling algorithms and memory-protection schemes.

The development of hardware and software for rapid transfer of data between core and secondary storage essentially removed the limits on the former by mapping it into the latter by segments, or "pages," and swapping them in and out as required by the program currently running. Such a system could then circulate control among a large number of programs, some or all of which could be processes interacting online with users at consoles (time-sharing). With each step in this development, applications programmers moved farther down an expanding hierarchy of layers of control that intervened between them and the computer itself. Only the layer at the top corresponded to a real machine; all the rest were virtual machines requiring translation to the layer above. Indeed, in IBM's OS/360 even the top layer was a virtual machine, translated by microprogrammed firmware into the specific instruction sets of the computers making up System/360. Despite appearances of direct control, this layering of abstract machines was as true of interactive systems as of batch systems. It remains true of current personal computing environments, the development of which has for the most part recapitulated the evolution of mainframe systems, adding to them a new layer of graphical user interfaces (GUIs). For example, Windows NT does not allow any application to communicate directly with the basic I/O system (BIOS), thus disabling some DOS and Windows 9x software.

What is important for present purposes about this highly condensed account of a history which remains largely uninvestigated is the extent to which operating systems increasingly realized the ideal of the computer as a self-programming device. In the evolution from the monitors of the mid-1950s to the interactive time-sharing systems of the early 1970s, programs themselves became dynamic entities. The programmer specified in abstract terms the structure of the data and the flow of computation. Making those terms concrete became the job of the system software, which in turn relied on increasingly elaborate addressing schemes to vary the specific links in response to run-time conditions. The operating system became the master choreographer in an ever-more complex dance of processes, coordinating them to move tightly among one another, singly and in groups, yet without colliding. The task required the development of sophisticated techniques of exception-handling and dynamic data management, but the possibility of carrying it out at all rested ultimately on the computer's capacity to rewrite its own tape.

Software Systems

Having concentrated during the 1960s on programming languages and operating systems as the means of addressing the problems of programming and software, the computing community shifted, or at least split its attention during the following decade. Participants at the 1968 NATO Conference on Software Engineering reinforced each other's growing sense that the cost overruns, slips in schedule, and failure to meet specifications that plagued the development of large-scale and mission-critical software systems reflected a systemic disorder, to be remedied only by placing "software manufacture . . . on the types of theoretical foundations and practical disciplines that are traditional in the established branches of engineering." Different views of the nature of engineering led to different approaches to this goal, but in general they built on developments in systems software, extending programming languages and systems to encompass programming methodologies. Two main strains are of particular interest here: the use of programming environments to constrain the behavior of programmers and the extension of programming systems to encompass and ultimately to automate the entire software development cycle.[6]

By the early 1970s, it seemed clear that, whatever the long-range prospects for automatic programming or at least for programming systems capable of representing large-scale computations in effective operational form, the development of software over the short term would rely on large numbers of programmers. Increasingly, programming systems came to be viewed in terms of disciplining programmers. Structured programming languages, enforced by diagnostic compilers, were aimed at constraining programmers to write clear, self-documenting, machine-independent programs. To place those programmers in a supportive environment, software engineers turned from mathematics and computer science to industrial engineering and project management for models of engineering practice. Arguing that "economical products of high quality are not possible (in most instances) when one instructs the programmer in good practice and merely hopes that he will make his invisible product according to those rules and standards," R. W. Bemer of GE spoke in 1968 of a "software factory" centered on the computer:

It appears that we have few specific environments (factory facilities) for the economical production of programs. I contend that the production costs are affected far more adversely by the absence of such an environment than by the absence of any tools in the environment (e.g., writing a program in PL/1 is using a tool).

A factory supplies power, work space, shipping and receiving, labor distribution, and financial controls, etc. Thus a software factory should be a programming environment residing upon and controlled by a computer. Program construction, checkout and usage should be done entirely within this environment. Ideally it should be impossible to produce programs exterior to this environment.[7]

Much of the effort in software engineering during the 1970s and 1980s was directed toward the design and implementation of such environments, as the concept of the "software factory" took on a succession of forms. CASE (computer-assisted software engineering) tools are perhaps the best example.

The Grail of Automatic Programming

While some software engineers thought of factories in terms of human workers organized toward efficient use of their labor, others looked to the automated factory first realized by Henry Ford's assembly line, where the product was built into the machines of production, leaving little or nothing to the skill of the worker. One aspect of that system attracted particular attention. Production by means of interchangeable parts was translated into such concepts as "mass-produced software components," modular programming, object-oriented programming, and reusable software. At the same time, in a manner similar to earlier work in compiler theory or indeed as an extension of it, research into formal methods of requirements analysis, specification, and design went hand in hand with the development of corresponding languages aimed at providing a continuous, automatic translation of a system from a description of its intended behavior to a working computer program. These efforts have so far met with only limited success. The production of programs remains in the hands of programmers.

CHAPTER SIX

Extracts from *The Roots of Software Engineering*

Editorial note: Mahoney presented an early version of what became "Finding a History for Software Engineering" in 1990 and published it as "The Roots of Software Engineering" in CWI Quarterly, the newsletter of the Dutch institute at which he made the presentation. Thanks to its inclusion on Mahoney's Web site, the paper was widely read. During its long gestation, the paper's framing and perspective evolved considerably and some interesting material from the earlier version was eliminated. I present two extracts from "The Roots of Software Engineering" here. The first is Mahoney's introduction, in which he gives a clear statement of his concern for precedents and connects this concern to the work of Thomas Kuhn. The second sketches the explosion of programming activity as computers entered commercial use during the 1950s and depicts as a response to this explosion the emergence of software as a recognized class of program.

Introductory Extract

At the International Conference on the History of Computing held in Los Alamos in 1976, R. W. Hamming placed his proposed agenda in the title of his paper: "We Would Know What They Thought When They Did It." He pleaded for a history of computing that pursued the contextual development of ideas, rather than merely listing names, dates, and places of "firsts." Moreover, he exhorted historians to go beyond the documents to "informed speculation" about the results of undocumented practice. What people actually did and what they thought they were doing may well not be accurately reflected in what they wrote and what they said they were thinking. His own experience had taught him that.[1]

Historians of science recognize in Hamming's point what they learned from Thomas Kuhn's *Structure of Scientific Revolutions* some time ago, namely, that the practice of science and the literature of science do not necessarily coincide. Paradigms (or, if you prefer with Kuhn, disciplinary matrices) direct not so much what scientists say as what they do. Hence, to determine the paradigms of past science historians must watch scientists at work practicing their science. We have to reconstruct what they thought from the evidence of what they did, and that work of reconstruction in the history of science has often involved a certain amount of speculation informed by historians' own experience of science. That is all the more the case in the history of technology, where up to the present century the inventor and engineer have—as Derek Price once put it—"thought with their fingertips," leaving the record of their thinking in the artifacts they have designed rather than in texts they have written.

Yet, on two counts, Hamming's point has special force for the history of computing. First, whatever the theoretical content of the subject, the main object of computing has been to do something, or rather to make the computer do something. Successful practice has been the prime measure of effective theory. Second, the computer embodies a historically unique relation of thinking and doing. It is the first machine for doing thinking. In the years following its creation and its introduction into the worlds of science, industry, and business, both the device and the activities involved in its use were new.

It is tempting to say they were unprecedented, were that not to beg the question at hand. Precedents are what people find in their past experience to guide their present action. Conversely, actions usually reflect the guidance of experience. Nothing is really unprecedented. Faced with a new situation, people liken it to familiar ones and shape their response on the basis of the perceived similarities. In the case of the computer, what was new was the reliable electronic circuitry that made its underlying theoretical structure realizable in practice. At heart, it was a Turing machine that operated within the constraints of real time and space. That much was unprecedented. Beyond that, precedent shaped the computer. The Turing machine was an open schema for a potentially infinite range of particular applications. How the computer was going to be used depended on the experience and expectations of the people who were going to use it or were going to design it for others to use.

As part of a history of the development of the computer industry from 1950 to 1970 focusing on the origins of the "software crisis," I am currently trying to determine what people had in mind when they first began to talk about "software engineering." Although one writer has suggested that the term originated in 1965, it first came into common currency in 1967 when the Study Group on Computer Science of the NATO Science Committee called for an international conference on the subject.[2]

Extract on the Growth of Programming and the Emergence of Software

The electronic digital stored-program computer marks the convergence of two essentially independent lines of development tracing back to the early nineteenth century, namely, the design of mechanical calculators capable of automatic operation and the development of mathematical logic. In outline, at least, those stories are reasonably well known and need no repetition here. Viewing them as convergent rather than coincident emphasizes that the computer emerged as the joint product of electrical engineering and theoretical mathematics and was shared by those two groups of practitioners, whose expertise intersected in the machine and overlapped on the instruction set. Both groups apparently looked upon programming as incidental to their respective concerns. Working with the model of the Turing machine, mathematical logicians concerned themselves with questions of computability considered independently of any particular device, while electrical engineers concentrated on the synthesis and optimization of switching circuits for specific inputs and outputs. Numerical analysts embraced the machine as part of their subject and hence took programming it as part of their task. B. V. Bowden of Ferranti, Ltd., editor of *Faster than Thought*, was unusual in even raising the question. As the number of computers in use in England in 1953 reached 150, he pointed to the growing difficulties and inefficiencies of programming and wondered where the programmers would come from.

> We have yet to analyse, for we have almost ignored them, the restrictions on machine performance which are due to the difficulties experienced by the operators who have to prepare programmes for them. It is significant that many machines have spent half their working lives in checking programmes and finding mistakes in them and only perhaps a third of the time in straightforward computation. One can deduce from this the startling conclusion that had the machines been a thousand times as fast as they are, their total output would not have been increased by more than about fifty per cent; in the last analysis the correction of programming errors depends almost entirely on the skill and speed of a mathematician, and there is no doubt that it is a very difficult and laborious operation to get a long programme right.[3]

As long as the computer remained essentially a scientific instrument, Bowden's concern found little echo; programming remained relatively unproblematic.

But the computer went commercial in the early 1950s. Why and how is another story. With commercialization came rapid strides in hardware—faster processors, larger memories, more efficient peripherals—together with equally rapid expansion of the imaginations of marketing departments. To sell the computer, they spoke not only of high-speed accounting, but of computer-based management. Again, at first few if any seemed concerned about who would write the programs needed to make it useful. IBM, for example, did not recognize "programmer" as a job category nor create a career track for it until the late 1950s.[4]

Companies soon learned that they had reduced the size of their accounting departments only to create ever-growing data-processing divisions, or to retain computer service organizations which themselves needed ever-more programmers. The process got under way in the late 1950s, and by 1968 some 500 companies were producing software. They and businesses dependent on them were employing some 100,000 programmers and advertising the need for 50,000 more. By 1970, the figure stood around 175,000. In this process, programs became "software" in two senses. First, a body of programs took shape (assemblers, monitors, compilers, operating systems, etc.) that transformed the raw machine into a tool for producing useful applications, such as data processing. Second, programs became the objects of production by people who were not scientists, mathematicians, or electrical engineers.

The increasing size of software projects introduced two new elements into programming: separation of design from implementation and management of programmers. The first raised the need for techniques for designing programs—often quite large programs—without writing them and for communicating designs to the programmers; the second, the need for means of measuring and controlling the quality of programmers' work. For all the successes of the 1960s, practitioners and managers generally agreed that those needs were not being met.

CHAPTER SEVEN

Finding a History for Software Engineering

ATING FROM the first international conference on the topic in October 1968, software engineering just turned thirty-five. It has all the hallmarks of an established discipline: societies (or subsocieties), journals, textbooks and curricula, even research institutes. Software Engineering would seem ready to have a history. Yet, a closer look at the field raises the question of just what the subject of the history would be. It is not hard to find definitions. A leading practitioner spoke of it in 1989 as "the disciplined application of engineering, scientific, and mathematical principles and methods to the economical production of quality software." But it is also not hard to find doubts about whether its current practice meets those criteria and, indeed, whether it is an engineering discipline at all. A colleague of the practitioner just quoted declared at about the same time (1990): "Software engineering is not yet a true engineering discipline, but it has the potential to become one." From the outset, software engineering conferences have routinely begun with a keynote address that asks "Are we there yet?" and proposes yet another specification of just where "where" might be.[1]

Because the field has been a moving target for its own practitioners, historians may understandably have trouble knowing just where to aim their attention. What is a history of software engineering about? Is it about the engineering of software? If so, by what criteria or model of engineering? Is it engineering as applied science? If so, what science is being applied and what is its history? Is it about engineering as project management? Is it engineering by analogy to one of the established fields of engineering? If so, which fields, and what are the terms of the analogy? Of what history would the history of software engineering be a part, that is, in what larger historical context does it most appropriately fit? Is it part of the history of engineering? The history of business and management? The history of the professions and of professionalization? The history of the disciplines and their formation? If several or all of these are appropriate, then what aspects of the history of software engineering fit where?

Alternatively, to put the question in another light, is the historical subject more accurately described as "software engineering" with the inverted commas as an essential part of the title? What seems clear from the literature of the field from its very inception, reinforced by addresses, panels, articles, and letters to the editor that continue to appear regularly, is that its practitioners disagree on what software engineering is, although most of them freely confess that, whatever it is, it is not (yet) an engineering discipline. Historians have no stake in the outcome of that question. They can just as readily write a history of "software engineering" viewed as the continuing effort of various groups engaged in the production of software to establish their practice as an engineering discipline. The question of interest to historians would then be how "software engineers" have tackled that task of self-definition. In large part, addressing that question comes down to observing and analyzing the answers practitioners have offered to the questions just posed. That is, rather than positing a consensus among practitioners concerning the nature of software engineering, historians can follow the efforts to achieve a consensus. Taking that approach would place the subject firmly in the comparative context of the history of professionalization and the formation of new disciplines.[2]

For this reason, it may help to think of historians and practitioners as engaged in a common pursuit. Both seek a history for software engineering, though not for the same purpose and not from the same standpoint. Hence, this chapter's title is meant to be ambiguous. In one sense, it describes historians trying to determine just what the subject of their inquiry might be and then deciding how to write its history. In another sense, it describes efforts by practitioners to define or to characterize software engineering. Often those efforts amount to finding a history, that is, seeking to identify the current development of software engineering with the historical development of one of the established engineering disciplines or of engineering itself. Using history in this way has its real dangers; the initial conditions cannot by their nature be exactly repeated. Nonetheless, it is essential both that one have the right history and that one have the history right, not least because what passes for history often amounts to common wisdom, folklore, or local myth. Here historians may offer some assistance to the software engineers. While we may not be able to tell them whether they have the right history, we can in many cases tell them what history they have chosen and whether they have got it right.[3]

Ultimately, every definition of software engineering presupposes some historical model. For example, take the oft-quoted passage from the introduction to the proceedings of the first Software Engineering Conference, convened by the NATO Science Committee in 1968: "The phrase 'software engineering' was deliberately chosen as being provocative, in implying the need for software manufacture to be based on the types of theoretical foundations and practical disciplines that are traditional in the established branches of engineering."[4]

The phrase indeed turned out to be provocative, if only because it left all the crucial terms undefined. What does it mean to "manufacture" software? Is that a goal or a current practice? What, precisely, are the "theoretical foundations and practical disciplines" that underpin the "established branches of engineering"? What roles did they play in the formation of the engineering disciplines? Is the story the same in each case? The reference to "traditional" makes the answer to that question a matter of history. It is a question of how the fields of engineering took their present form. It is a search for historical precedents, or what we have come to refer to as "roots."

Or rather, it is a matter of what I call "agendas." The agenda of a field consists of what its practitioners agree ought to be done, a consensus concerning the problems of the field, their order of importance or priority, the means of solving them (the tools of the trade), and perhaps most importantly, what constitutes a solution. Becoming a recognized practitioner means learning the agenda and then helping to carry it out. Knowing what questions to ask is the mark of a full-fledged practitioner, as is the capacity to distinguish between trivial and profound problems; "profound" means moving the agenda forward. One acquires standing in the field by solving the problems with high priority, and especially by doing so in a way that extends or reshapes the agenda, or by posing profound problems. The standing of the field may be measured by its capacity to set its own agenda. New disciplines emerge by acquiring that autonomy. Conflicts within a discipline often come down to disagreements over the agenda: what are the really important problems?[5]

A new science means a new agenda, and tracing the emergence of a new science means showing how a group of practitioners coalesces around a common agenda different from other agendas in which they have been engaged. Each of those other agendas reflects a history, and so the members of the group bring to their new agenda a variety of histories. Some, or perhaps even much, of the disagreement among the participants in the first two NATO conferences, especially the second, rested on the different histories they brought to the gatherings. None of them was a software engineer, for the field did not exist. Rather, people came from quite varied professional and disciplinary traditions, each of which had its own history, in many cases a mythic history. What follows is a brief look at how the histories have been invoked and how they have been understood.[6]

Models of Engineering: Historical Precedents

Three histories in particular have directed the practitioners' search for historical guidance: applied science, mechanical engineering, and industrial engineering and management.

Applied Science

To some, in particular many of the European participants, engineering was essentially applied science, and the science in question here was mathematics. What was needed, then, was firm grounding in theoretical, i.e., mathematical, computer science. The historical model seemed clear. Indeed, it had been set forth explicitly almost ten years earlier, albeit in another context, by John McCarthy, the creator of LISP and co-founder of artificial intelligence. Looking "towards a mathematical theory of computation" at IFIP 1962, he had reached for a familiar touchstone: "In a mathematical science, it is possible to deduce from the basic assumptions, the important properties of the entities treated by the science. Thus, from Newton's law of gravitation and his laws of motion, one can deduce that the planetary orbits obey Kepler's laws."[7]

As McCarthy and his audience well knew, one can also deduce the laws of the motion of terrestrial bodies and all the mechanics that derives from them. He extended the model at the conclusion of his 1963 article, "A Basis for a Mathematical Theory of Computation," by reference to later successes in mathematical physics: "It is reasonable to hope that the relationship between computation and mathematical logic will be as fruitful in the next century as that between analysis and physics in the last. The development of this relationship demands a concern for both applications and mathematical elegance."[8]

The applications of mathematics to physics had produced more than new theories. The mathematical theories of thermodynamics and electricity and magnetism had informed the development of heat engines, of dynamos and motors, of telegraphy and radio. Those theories formed the scientific basis of engineering in those fields.

The twentieth century had a new science, McCarthy believed, and it too had implications beyond just theory. "Computation is sure to become one of the most important of the sciences," he began:

> This is because it is the science of how machines can be made to carry out intellectual processes. We know that any intellectual process that can be carried out mechanically can be performed by a general purpose digital computer. Moreover, the limitations on what we have been able to make computers do so far clearly come far more from our weakness as programmers than from the intrinsic limitations of the machines. We hope that these limitations can be greatly reduced by developing a mathematical science of computation.[9]

The ultimate object of computer science was working programs, argued McCarthy, and a suitable theory of computation would provide:

> first, a universal programming language along the lines of Algol but with richer data descriptions; second, a theory of the equivalence of computational processes, by which equivalence-preserving transformations would allow a choice of among various forms of an algorithm, adapted to particular circumstances; third, a form of symbolic representation of algorithms that could accommodate significant changes in behavior by simple changes in the symbolic expressions; fourth, a formal way of representing computers along with computation; and finally a quantitative theory of computation along the lines of Claude Shannon's measure of information.[10]

Note that as this list progresses, it sounds more and more like engineering, and McCarthy's agenda (and its history) continued to echo in the software-engineering literature. In arguing in 1984 that "professional programming practice should be based on underlying mathematical theories and follow the traditions of better-established engineering disciplines," C. A. R. Hoare highlighted in a sidebar McCarthy's comparison of physics and mathematical logic quoted above.[11]

Over the decade of the 1960s theoretical computer science achieved standing as a discipline recognized by both the mathematical and the computing communities, and it could point to both applications and mathematical elegance. Yet, it took the form more of a family of loosely related research agendas than of a coherent general theory validated by empirical results. No one mathematical model had proved adequate to the diversity of computing, and the different models were not related in any effective way. What mathematics one used depended on what questions one was asking, and for some questions no mathematics could account in theory for what computing was accomplishing in practice. It was a far cry from Newton's mechanics, much less the mathematical physics of the nineteenth century, and it remains so.[12]

At the second NATO Conference on Software Engineering held in Rome in October 1969, Christopher Strachey, director of the Programming Research Group at Oxford University and a leading figure in the development of formal semantics, lamented that "one of the difficulties about computing science at the moment is that it can't demonstrate any of the things that it has in mind; it can't demonstrate to the software engineering people on a sufficiently large scale that what it is doing is of interest or importance to them."[13]

About a decade later, a committee in the United States reviewing the state of art in theoretical computer science echoed his diagnosis, noting the still limited application of theory to practice. By the mid-1970s, moreover, it seemed clear to some that, even if existing theory had practical application, it would not quite meet the needs of software engineering. In a 1976 article, Barry Boehm of TRW proposed that software engineering be defined as "the practical application of scientific knowledge in the design and construction of computer programs and the associated documentation required to develop, operate, and maintain them." Boehm identified the salient terms as "design," "software maintenance," and "scientific knowledge" and took stock of what was known in each area.[14]

The first two terms he addressed by reference to what by then was becoming the standard model of the "software life cycle," a sequence that took a project from the requirements to an operating program by way of specification, design, coding, and testing. What he saw as current practice reinforced the concerns of the crisis. In particular, requirements analysis was informal at best, and software design was "still almost completely a manual process . . . [with] relatively little effort devoted to design validation and risk analysis." Yet, as he had shown in a now classic article in 1973, the bulk of the errors in software were made during the design phase.[15]

Most significantly for present purposes, he also concluded that little of current computer science was relevant to the problems of software engineering:

> Those scientific principles available to support software engineering address problems in an area we shall call *Area 1: detailed design and coding* of *systems software* by *experts* in a relatively *economics-independent* context. Unfortunately, the most pressing software development problems are in an area we shall call *Area 2: requirements analysis design, text, and maintenance* of *applications software* by *technicians* in an *economics-driven* context.[16]

However successful the experimental systems and theoretical advances produced in the laboratory, especially the academic laboratory, they did not take account of the challenges and constraints of "industrial-strength" software in a competitive market. As Fritz Bauer, the organizer of the first NATO conference, had put it at IFIP '71, those problems made software engineering "the part of computer science that is too difficult for the computer scientists."[17]

Mechanical Engineering

If not applied science, then what? Others at the NATO conference had proposed models of engineering that emphasized analogies of practice rather than theory. Perhaps the most famous of these was M. D. McIlroy's evocation of the machine-building origins of mechanical engineering and the system of mass production by interchangeable parts that grew out of them. Seeing software sitting somewhere on the other side of the Industrial Revolution, he proposed to vault it into the modern era.

> We undoubtedly produce software by backward techniques. We undoubtedly get the short end of the stick in confrontations with hardware people because they are the industrialists and we are the crofters. Software production today appears in the scale of industrialization somewhere below the more backward construction industries. I think its proper place is considerably higher, and would like to investigate the prospects for mass-production techniques in software.

He left no doubt of whose lead to follow. He continued:

> In the phrase "mass production techniques," my emphasis is on "techniques" and not on mass production plain. Of course mass production, in the sense of limitless replication of prototype, is trivial for software. But certain ideas from industrial technique I claim are relevant. The idea of subassemblies carries over directly and is well exploited. The idea of interchangeable parts corresponds roughly to our term "modularity," and is fitfully respected. The idea of machine tools has an analogue in assembly programs and compilers. Yet this fragile analogy is belied when we seek for analogues of other tangible symbols of mass production. There do not exist manufacturers of standard parts, much less catalogues of standard parts. One may not order parts to individual specifications or size, ruggedness, speed, capacity, precision or character set.[18]

As studies of the American machine-tool industry during the nineteenth and early twentieth centuries have shown, McIlroy could hardly have chosen a more potent model (he has a long-standing interest in the history of technology). Between roughly 1820 and 1880, developments in machine-tool technology had increased routine shop precision from .01" to .0001". More importantly, in a process characterized by the economist Nathan Rosenberg as "convergence," machine-tool manufacturers learned how to translate new techniques developed for specific customers into generic tools of their own. So, for example, the need to machine bits for drilling small holes in percussion locks led to the development of the vertical turret lathe, which in turn lent itself to the production of screws and small precision parts, which in turn led to the automatic turret lathe. Indeed, it was precisely the automatic screw-cutting machine that McIlroy had in mind.[19]

As McIlroy noted, he was giving sharper, historically grounded form to an idea that had already begun to take shape. In an Advanced Course in Software Engineering that took place at Munich's Technical University in 1972, Jack B. Dennis of MIT's Project MAC lectured on "Modularity," pointing as example to standardized floor tiles (19" square "modules") which fill any size or shape of floor area "with just a bit of trimming at the boundary," while allowing great variety through different colors and textures of modules.

> In modular software, clearly the "standardized units or dimensions" should be standards such that software modules meeting the standards may be conveniently fitted together (without "trimming") to realize large software systems. The reference to "variety of use" should mean that the range of module types available should be sufficient for the construction of a usefully large class of programs.[20]

Especially as expressed by McIlroy, the idea has had a long career in software engineering. During the 1970s it directed attention beyond the development of libraries of subroutines to the notion of "reusable" programs across systems, and in the 1980s it underlay the growing emphasis on object-oriented programming as the means of achieving such reusability on a broad scale. Modularity is essentially what Brad Cox was looking for around 1990 as the basis for software's "industrial revolution." More generally, the analogy with machine building and the metaphorical language of machine-based production became a continuing theme of software engineering, often illustrated by pictures of automobile assembly lines, as in the case of Peter Wegner's four-part article in *IEEE Software* in 1984 on "Capital-Intensive Software Technology." The cover of that issue bore a photograph of a Ford assembly line in the 1930s, and a picture of the same line in the early 1950s adorned Gregory W. Jones's *Software Engineering*.[21]

Industrial Engineering

As the move from machine tools to the assembly line suggests, McIlroy's model of mechanical engineering was closely akin to F. L. Bauer's proposal at IFIP 71 that "software design and production [be viewed] as an industrial engineering field."

> For the time being, we have to work under the existing conditions, and the work has to be done with programmers who are not likely to be re-educated. It is therefore all the more important to use organizational and managerial tools that are appropriate to the task.[22]

On that model the problems of large software projects came down to the "division of the task into manageable parts," its "division into distinct stages of development," "computerized surveillance," and "management." Each of these tasks posed significant problems, and Bauer had specific suggestions to make only with regard to the third: computerized surveillance consisted of:

> Automatic updating and quality control of documentation,
> Selective dissemination of information to all project staff,
> Surveillance of deadline plans,
> Collection of data for simulation studies,
> Collection of data for quality control,
> Automatic production of manuals and maintenance material.

"It is clear," he noted, "that a house well equipped with programs and an underlying philosophy for doing these things, can be regarded as a modern software plant."[23]

Bauer's idea was not new. In a "Position Paper for Panel Discussion [on] the Economics of Program Production" at IFIP 68, also presented in substance at the NATO conference, R. W. Bemer of GE had suggested that what software managers lacked was a proper environment:

> It appears that we have few specific environments (factory facilities) for the economical production of programs. I contend that the production costs are affected far more adversely by the absence of such an environment than by the absence of any tools in the environment (e.g., writing a program in PL/1 is using a tool).
>
> A factory supplies power, work space, shipping and receiving, labor distribution, and financial controls, etc. Thus a software factory should be a programming environment residing upon and controlled by a computer. Program construction, checkout, and usage should be done entirely within this environment. Ideally it should be impossible to produce programs exterior to this environment.[24]

Bemer's proposal was aimed at the problem of workers' near-total control over production, which the computer itself held promise of overcoming. "Economical products of high quality," he continued,

> are not possible (in most instances) when one instructs the programmer in good practice and merely hopes that he will make his invisible product according to those rules and standards. This just does not happen under human supervision.
>
> A factory, however, has more than human supervision. It has measures and controls for productivity and quality. Financial records are kept for costing and scheduling. Thus management is able to estimate from previous data: not so with programming management in general. Computer supervision and aid are vital, with the accent upon human engineering factors so that working in the environment is both attractive and effective for the programmer.

In reading these words, it is hard not to hear an echo of Frederick W. Taylor and his methods of "scientific management," which informed management thinking, both here and in Europe in ways that are only now becoming clear.[25] Indeed, the basic principles of Taylor's system sound much like the agenda that early software engineer-managers were laying out for themselves. The primary obligation of management according to Taylor was to determine the scientific basis of the task to be accomplished. That came down to four main duties:

> *First*. They develop a science for each element of a man's work, which replaces the old rule-of-thumb method.
>
> *Second*. They scientifically select and then train, teach, and develop the workman, whereas in the past he chose his own work and trained himself as best he could.
>
> *Third*. They heartily cooperate with the men so as to insure all of the work [is] being done in accordance with the principle of the science which has been developed.
>
> *Fourth*. There is an almost equal division of the work and the responsibility between the management and the workmen. The management take over all work for which they are better fitted than the workmen, while in the past almost all of the work and the greater part of the responsibility were thrown upon the men.[26]

In the emphasis on supervision and support of the programmer, Bemer's factory sounds like Taylor's machine shop, with management seeking to impose the "one best way" over a worker still in control of the shop floor.

A decade later, William W. Agresti of the University of Michigan-Dearborn made the tie to Taylor explicit. After his talk, "Applying Industrial Engineering to the Software Development Process," presented at the

IEEE Computer Society's 1981 International Conference, he published a follow-up piece:

> While working on this project, I returned for inspiration to the "old masters" of industrial engineering: Frederick Taylor, Henry Gantt, and Frank and Lillian Gilbreth. The accounts of their work in the early 1900s provide remarkable reading as a glimpse of society at that time. I was also impressed that much of what they were saying then about I.E. (or "scientific management" as it was known then) could be said today about software engineering.[27]

As examples, Agresti offered a page of excerpts from the works of the masters as they might apply to such matter as "Finding Program 'Bugs,'" "Introducing Structured Programming Methods," and "Software Tools." Concerning the "Analysis of Algorithms," he went to the heart of Taylor's system: "Now, among the various methods used . . . there is always one method which is quicker and better than any of the rest. And this one best method can only be discovered through a scientific study and analysis of all the methods in use."

Whether implicitly or explicitly, Taylorism continued to inform the industrial approach to software engineering. Leon J. Osterweil's keynote address at the Ninth International Conference on Software Engineering in 1987 offers a striking example. Even more recently, Watts S. Humphrey, principal designer of the widely used (and DoD-sanctioned) Capability Maturity Model and Personal Software Process, provides more explicit testimony to Taylor's presence in thinking about software management. In an article on the current status and trends in the Personal Software Process, Humphrey references Peter Drucker in asserting that "Even though manual and intellectual tasks are significantly different, we can measure, analyze, and optimize both and thus apply Taylor's principles equally well." He then explains his point in language quite close to Bemer's:

> The principal difference between manual and intellectual work is that the knowledge worker is essentially autonomous. That is, in addition to deciding how to do tasks, he or she must also decide what tasks to do and the order in which to do them. The manual worker commonly follows a relatively fixed task order, essentially prescribed by the production line. So studying and improving the performance of intellectual work must not only address the most efficient way to do each task but also consider how to select and order these tasks. This is essentially the role of a defined process and a detailed plan. The process defines the tasks, task order, and task measures, while the plan sizes the tasks and defines the task schedule for the job being done.[28]

If, as Osterweil maintains, "software processes are software, too," then some of those processes are Taylor-inspired software for managing workers.

Yet, in the late 1960s, when the notions of software engineering and the software factory were first proposed, practitioners could fulfill none of Taylor's requirements. To what extent computer science could replace rule of thumb in the production of software was precisely the point at issue at the NATO conferences (and, as noted above, it remains a question). Even the optimists agreed that progress had been slow. Unable, then, to fulfill the first duty, programming managers were hardly in a position to carry out the third. Everyone bemoaned the lack of standards for the quality of software. As far as the fourth was concerned, few ventured to say who was best suited to do what in large-scale programming projects. Frederick P. Brooks offered an example in his now classic *The Mythical Man-Month*:

> It is a very humbling experience to make a multimillion-dollar mistake, but it is also very memorable. I vividly recall the night we decided how to organize the actual writing of external specifications for OS/360. The manager of architecture, the manager of control program implementation, and I were threshing out the plan, schedule, and division of responsibilities.
>
> The architecture manager had 10 good men. He asserted that they could write the specifications and do it right. It would take ten months, three more than the schedule allowed.
>
> The control program manager had 150 men. He asserted that they could prepare the specifications, with the architecture team coordinating; it would be well-done and practical, and he could do it on schedule. Furthermore, if the architecture team did it, his 150 men would sit twiddling their thumbs for ten months.
>
> To this the architecture manager responded that if I gave the control program team the responsibility, the result would *not* in fact be on time, but would also be three months late, and of much lower quality. I did, and it was. He was right on both counts. Moreover, the lack of conceptual integrity made the system far more costly to build and change, and I would estimate that it added a year to debugging time.[29]

Only with that experience behind him was Brooks in a position to think about what precisely was wrong with his decision.

As for Taylor's second principle, by 1969 the failure of management to establish standards for the selection and training of programmers was legend. As Dick H. Brandon, the head of one of the more successful software houses, pointed out, the industry at large scarcely agreed on the most general specifications of the programmer's task. Forced to hire people without programming experience managers had only one (dubious) aptitude test at their disposal, and no one knew for certain how to train those people once hired. So one was back where one started: to implement the model required solving the problems to which the model was supposed to provide the solution, quite apart from how effective that solution had in fact turned out to be.[30]

Much of the articulation of software engineering during the 1970s and 1980s aimed at laying the groundwork for effective management: structured analysis and design as a means of hierarchical division of projects and allocation of tasks, structured programming as a means both of quality control and of disciplining programmers, methods of cost accounting and estimation, methods of verification and validation, techniques of quality assurance. Except for structured programming, which could be enforced by increasingly effective diagnostic compilers, most of these methods were paper exercises for which the computer served largely clerical purposes. One could very well program "outside the environment."

A year after Bemer laid out his scheme, GE left the computer business, but the concept of the software factory survived. Indeed, the Systems Development Corporation trademarked the term, and proposed to set up what Michael Cusumano describes as a "conveyor and control system that brought work and materials (documents, code modules) through different phases, with workers using standardized tools and methods to build finished software products," or, in the words of its designers,

> In the Factory, the Development Data Base serves as the assembly line—carrying the evolving system through the production phases in which factory tools and techniques are used to steadily add more and more detail to the system framework.[31]

The evocation of the assembly line linked the software factory to a model of industrial production different from Taylor's—how different is a complex historical and technical question—namely Ford's system of mass production through automation. Ford did not have to concern himself about how to constrain workers to do things in "the one best way." His machines of production embodied that way of doing things; the worker had little to do with it. The same was true of the assembly line itself.

> In the chassis assembling are forty-five separate operations or stations. The first men fasten four mud-guard brackets to the chassis frame; the motor arrives on the tenth operation and so on in detail. Some men do only one or two small operations, others do more. The man who places a part does not fasten it—the part may not be fully in place until after several operations later. The man who puts in a bolt does not put on the nut; the man who puts on the nut does not tighten it.[32]

As parts moved through the production process, they took on the shape of the Model T because that shape was, so to speak, built into the machines of production. Ford's methods worked because he was producing a machine, the essential components of which could be completely and precisely specified and hence could be produced by machines, themselves in turn fully specifiable. Indeed, Ford designed the Model T to be produced by machines, and therefore the available means of production were part of the target specifications. Underpinning that achievement was the development of the machine-tool industry alluded to above.

The assembly line has held continuing allure for software engineers, who generally find it ironic that "programmers have done a good job of automating everyone's work but their own." Indeed, that is referred to as the "software paradox." That one would find it paradoxical lies in the nature of the computer combined with a particular notion of engineering. "We know," said John McCarthy, "that any intellectual process that can be carried out mechanically can be performed by a general purpose digital computer." By "mechanically," he meant according to clear, unambiguous procedures. Engineering, especially science-based engineering, aims at providing solutions of just that sort to its problems. Hence, one ought to be able to do for software what one has done for other engineering problems, namely, to transfer solutions to the computer for execution. The grail of "automatic programming," as pursued in particular by Robert Balzer of ISI, with the support of DARPA, throughout the 1970s and 1980s was a software development system which could take a problem specification and transform it automatically into a working system as solution, in essence eliminating the programmer. Much of the CASE software developed during this period purported to achieve portions of that goal; on close inspection, little of it in fact lived up to its claims.[33]

In *Japan's Software Factories* and related articles, Michael Cusumano presents evidence suggesting that the effective management of software production lies at a level in between craft production and mass production, namely, at the level of flexible design and production systems. The "factories" Cusumano examined during the mid-to-late 1980s involved the following measures:

- identification of a target market and of a range of "semi-custom" products for it
- a long-term commitment to production for that market
- intensive review of currently available tools and practices
- intensive and continuing training of personnel and imposition of a programming discipline on them
- commitment of productive effort to the building of tools
- emphasis on reusability, encouraging designers and programmers to devote project effort to non-project goals
- emphasis on design and testing phases of development
- intensive quality control through inspection and testing.

Basically, the list comes down to a corporate investment in training and maintaining a skilled workforce with cumulative experience in the areas for which the workers are building systems. Until recently, Bell Labs was perhaps the foremost example of such environments in the United States. UNIX is a leading model for the notion of software tools and of a "programmer's workbench."

To what extent such environments are "factories" in the sense in which they were originally conceived is debatable. Indeed, Cusumano notes the resistance of the programmers themselves to the term, because it connotes a devaluation of their skills. The phrase "workbench," which also appears in the Japanese context, lies closer to the shop than to the assembly line. Although these environments suggest that the production of software is far from inherently unmanageable, they also make clear that productivity depends on a highly skilled, and commensurately expensive workforce.

What Is Being Automated?

Those "factories" are also not likely to hold a solution for the problems of software production that motivated the drive for software engineering, but, then, neither is automatic programming. Consider another version of development phases of the software life cycle (fig. 5). We are on firmest theoretical ground at the bottom of the diagram. That is where computer science has achieved its most profound results, and that is where theory has most effectively translated into practical software tools. But the problems of air traffic control systems, of national weather systems, of airline booking systems, all lie at the top of the diagram, where a real-world system must be transformed into a computational model. That is where software engineering is not about software, indeed where it may not be about engineering at all.

Software engineering began as a search for an engineering discipline on which to model the design and production of software. That the search continues after thirty-five years suggests that software may be fundamentally different from any of the artifacts or processes that have been the object of traditional branches of engineering: it is not like machines, it is not like masonry structures, it is not like chemical processes, it is not like electric circuits or semiconductors. It thereby raises the question of how much guidance one may expect from trying to emulate the patterns of development of those engineering disciplines. During general discussion concerning theory and practice held on the last day of the Rome conference in 1969, I. P. Sharp came at the issue from an entirely different angle, arguing that one ought to think in terms of "software architecture" (design), which would be the meeting ground for theory (computer science) and practice (software engi-

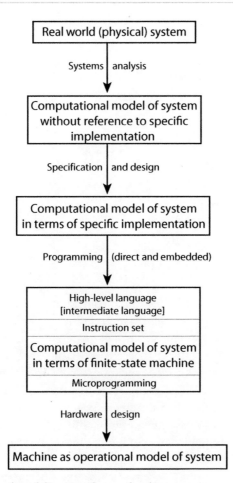

Figure 5   Levels of modeling in software development.

neering). "Architecture is different from engineering," he maintained and then added, "I don't believe for instance that the majority of what [Edsger] Dijkstra does is theory—I believe that in time we will probably refer to the 'Dijkstra School of Architecture.'" That is no small distinction. Architecture has a different history from engineering, and we train architects differently from engineers. It is striking to a historian looking for a history of software engineering that the Ninth Foundations of Software Engineering Conference in 1998, which concluded with a plenary session on whether software engineering is ready to become a "profession," that is, whether its practitioners should be subject to licensing as professional engineers, was preceded by the Third International Workshop on Software Architecture.[34]

CHAPTER EIGHT

Boys' Toys and Women's Work:
Feminism Engages Software

N HER CONTRIBUTION to the volume in which this chapter originally appeared, Pamela Mack reflected on her inability to find a
case in which feminism, particularly feminist theory, has improved
practice or made for better engineering.[1] What follows is an extended version of a commentary meant at the time to take up her implicit challenge
by considering a future case study of a situation still in flux, one in which
feminism might still make a difference in ways we can now specify. The
field in question is software engineering, but before turning to it I'd like
to consider the computer and computing more generally as a case study
of gender and technology.

As a case study, computing has the advantage of having emerged as an
essentially new technology at a definable time and place. Before the creation of the electronic digital computer in the mid-1940s, what we now
understand as computing did not exist. Before then, "computer" denoted
a person, usually a woman, who carried out calculations by hand or with
a mechanical calculator. The work was viewed as clerical: computers followed explicit instructions (usually provided by a male mathematician),
and accuracy took precedence over imagination. Indeed, when Turing
wanted an example of a "mechanical" procedure, he invoked the human
computer, albeit in masculine form.

Perhaps for that reason, the first programmers for the ENIAC were
women, as were several of the leading figures in the early development of
computing, most notably Grace Hopper. Yet, men soon took over, and leading women became the exception in an increasingly masculine field. Hopper remained highly visible until her death in 1992. Indeed, "Amazing

Grace" achieved something akin to canonization in her own lifetime. Everyone of a certain generation and older has a Hopper tale to tell, and the telling often has a hagiographical tone. Although Jean Sammet did not reach that exalted state of matron saint, she also remained a formidable figure within IBM and the profession, universally recognized for her encyclopedic knowledge of programming languages. One can name others, several of whom have served as president of the Association of Computing Machinery (ACM), but that is perhaps the point: one can name them.[2]

How and why did the field fall to men? Does the answer go beyond the obvious, namely, that in computing, as in other fields of war work, women returned to the home, or were directed back to the home, to make room for men returning to the labor force? The women of ENIAC, the first programmers, tell a common story of careers interrupted or ended by marriage and the arrival of children. Or is it a matter of men discovering that programming, unlike calculating, was a challenging, creative intellectual enterprise that promised rewards and reputation? At one of the first conferences on the history of computing, John Backus described the culture of the early programming community:

> Programming in the America of the 1950s had a vital frontier enthusiasm virtually untainted by either the scholarship or the stuffiness of academia. The programmer-inventors of the early 1950s were too impatient to hoard an idea until it could be fully developed and a paper written. They wanted to convince others. Action, progress, and outdoing one's rivals were more important than mere authorship of a paper. Recognition in the small programming fraternity was more likely to be accorded for a colorful personality, an extraordinary feat of coding, or the ability to hold a lot of liquor well than it was for an intellectual insight. Ideas flowed freely along with the liquor at innumerable meetings, as well as in sober private discussions and informally distributed papers.[3]

In short, programming quickly became a hard-drinking boys' club; the use of "fraternity" is revealing. By the late 1950s, men appear to have had the field to themselves. The groups that gathered informally at the RAND Corporation a day prior to the annual Western Joint Computer Conference to discuss the state of the profession included no women.

The (Trans)Gendering of the Computer

As the field became masculine, so too did the machine. At first, that may not seem puzzling. Given the traditional gendering of machines, what else would one expect of a room full of vacuum tubes and oscilloscopes, even as the tubes disappeared behind metal doors and the oscilloscope gave way to the video display? Yet, two questions immediately arise. First, by the early 1950s programming was emerging as an activity distinct from that concerned with the computer as a physical device. The split was reinforced with the development in the mid- to late 1950s of programming languages and operating systems, both designed to keep the programmer away from the machine. In most computing installations, the computer had a room of its own, in which it was tended by designated operators, often women, who mediated between it and the programmers. Academically and profession- ally, computer engineering took charge of the hardware, while computer science concerned itself with the software, that is, with programming lan- guages, operating systems, and methods of data management. Hence, while continuing cultural icons might explain the masculine character of the computer itself, they do not explain why programming should have been masculinized. The boys' club is not an explanation; it needs explaining.[4]

Second, the personal computer assumed a shape in the late 1970s and early 1980s that tied it to cultural icons and activities that in and of them- selves seem less straightforwardly masculine. The video screen, keyboard, and mouse invited users to write, to draw, to play games, to communi- cate. Indeed, the emphasis on the verbal, the visual, and the conversa- tional would seem to have pushed the device more toward the feminine, but apparently, that is not how the culture has viewed it. Indeed, casting the computer in a quite different form does not seem to have changed things that much. Studies since the early 1980s have generally found that children of both sexes from kindergarten on identify the personal com- puter as masculine: it is something for the boys. The researchers them- selves have found that puzzling, but it may at least help to explain why the fundamental restructuring of the computer industry reflected in the replacement of IBM by Microsoft as the shaping force has apparently done little to alter cultural perceptions of the technology. The "triumph of the nerds" does not seem a victory for women.[5]

The central question seems clear. Computing is readily available public knowledge: anyone can acquire it. The machine is for sale everywhere: anyone can buy it. Any woman who wished to do so could become a com- puter scientist, computer engineer, or software developer. Yet, by any measure, the world of computing includes relatively few women. The reason seems at first equally clear. The door may be open, but the world beyond it does not invite entry. Computing is a masculine world, in which women do not feel comfortable. Yet, that is a restatement of the question, not an answer to it. It does not explain what it means for computing to be masculine and how it became so. Nor does it suggest what a regender- ing of the world of computing might entail and how it might be accomplished. By the same token, it does offer a first-order reply to the question that shapes this volume: what difference has feminism made? To judge by the literature on women and computing, none. But that is a first-order reply, and it requires some adjustment.[6]

Indeed, when one examines the question closely, it fragments into pieces that grow more puzzling as they become more specific. Are we talking about computer science or computer technology? In the latter case, are we talking about the technology of production or about computers out there in "Consumption Junction"? It's evident that women have not played, and still do not play, much of a role in computer science or in the design and development of computer systems, whether mainframe, minicomputer, or personal computer. They do seem to be more proportionately represented among the consumers of computing, where their representation would seem to be determined more by their presence in business and the professions than by a particular stance toward computers. But there too computing seems to undermine that presence by making women and their work invisible. All that seems clear. The problem is to explain why and to do so without naturalizing technology or essentializing gender.[7]

Computing lends itself to the challenge, since the computer as a concept is protean. In principle, a computer can do anything we can describe in certain ways that are limited more in theory than in practice. So in practice computing is what we have made of it since the creation of the computer in the late 1940s and early 1950s. Whatever one wants to say about such abstractions as the Turing machine, it is hard to know how physical computers and the systems running on them could be anything other than socially constructed. Computing has no nature. It is what it is because people have made it so.[8]

Those people have been overwhelmingly men. That is a readily verifiable fact. What they have created is overwhelmingly masculine. That is a readily verifiable perception, which one often reads in feminist explanations of the paucity of women in the field. Yet, although writers on the subject of gender and computing all proceed on the premise that computing is gendered masculine—it must be, else why would it be dominated by men?—the more closely they look at it, the less clear it is just what makes it so. As Judy Wajcman and others have pointed out, some of those explanations verge on the essentialist, as "masculine" and "feminine" take the pole positions in a set of dichotomies: abstract/concrete, objectivity/subjectivity, logical/intuitive, mind/body, dominating/submissive, war/peace, and so on. Computing generally winds up on the masculine side of each of them. But essentialism is not the only problem here. Under such an interpretation, a feminine form of computing becomes the complete obverse of what now exists, and that seems as unimaginable as it is unrealistic. Fortunately, a closer look, both at the dichotomies and at their application to computing, suggests that the solution is not that drastic. To show how, we need first to consider the ways in which efforts to pin down the masculine nature of computing end up blurring the very lines they try to draw. Let us work with three case studies from the recent literature.[9]

Crossing Boundaries

In "Women's Studies and Computer Science: Their Intersection," Thelma Estrin points to convergences between feminist epistemology on the one hand and developments in computing and cognitive science on the other. Aimed at reconciling two disciplines that "both emerged as academic disciplines in the 1960s, with no interaction between them because they evolved along very different paths—one for women and one for men," her discussion tends to cloud the grounds of their convergence rather than shedding light on them. For, without evident input from women or feminism, a masculine computer science seems by Estrin's own description to have developed precisely the modes of thinking that she and others want to identify as feminine.[10]

For example, asserting that "feminist epistemology supports concrete thinking as a valuable tool in our way of thinking," Estrin makes much of object-oriented programming as a style of thinking in concrete terms. Contrasted with procedural and functional programming, "object-oriented programming is regarded as a physical model simulating the behavior of either a real or imaginary part of the world (C++). Physical models have elements that directly reflect phenomena and concepts that undermine the canonical position by supporting trends that challenge established methods and encourage working with specific objects." The "(C++)" refers to perhaps the most popular object-oriented programming language in use today, which Estrin interestingly contrasts with C and with LISP as examples of the other modes of programming. The contrast is interesting for two reasons. First, C++ began life as a preprocessor for C; it was originally "C with Classes" and based on Simula, a simulation language dating back to the 1960s. It was created in exactly the same environment as C, namely the Computing Research department at Bell Labs, Murray Hill. Second, LISP was initially designed to facilitate just the sort of interactive programming that Estrin (following Sherry Turkle and Seymour Papert) identifies with "bricolage," and it was the basis for Logo, the children's language that Estrin praises highly for encouraging thinking about objects and their relations. Both LISP and Logo are products of MIT's AI Laboratory, not particularly renowned for its feminist outlook. Moreover, LISP today comes in flavors that include objects, which are conceptually quite compatible with it.[11]

If one considers the history of object-oriented programming in terms of its genealogy, then its proposed empathy with feminist modes of thinking appears all the more problematic. Estrin points specifically to Smalltalk, the language of Alan Kay's Dynabook, aimed at placing the power of the computer in the hands of children for drawing, writing, and creative programming. But, like C++, Smalltalk itself is also an offspring of Simula, which was designed not for children but for researchers doing systems analysis and simulations at the Norwegian Computing Center in the early 1960s. Within that genealogy, object-oriented programming emerges from the notion of modularity, or the division of a program into subunits which are independent of one another and of any specific context. An early and well-known expression of the notion is M. D. McIlroy's proposal in 1968 for "Mass-Produced Software Components," which is filled with analogies to machine tools and redolent with the smell of the machine shop and indeed the assembly line.[12]

Although one might think of object-oriented programming as rooted in the concrete, object-oriented design involves a sustained process of abstraction, in which one seeks to characterize the objects as members of classes, which themselves are elements of yet more general classes, so that objects may inherit the properties they share with members of other classes. Thus the object "square" is a kind of "rectangle," which in turn is a kind of "figure," and so on. The result is a hierarchy of classes, another concept not generally associated with the feminine. Deciding on what will be the objects constituting a program and how they are related with one another seems a quintessentially Aristotelian exercise in definition by abstraction. To the extent that abstract, hierarchical thinking is classified as masculine, it is difficult to see how object-oriented design and programming are any less so.

If, then, there is convergence between computer science and women's studies, it does not seem to result from the reshaping of computational thinking by feminist epistemology. What Estrin identifies as common elements have emerged within computing from what would seem to be masculine foundations. If the elements to which she points open computer science to women by making it amenable to their ways of thinking about the world, then it has been open for a long time, at least back to the 1960s, because they are ways of thinking to which the men who created computing themselves aspired.

Tove Håpnes's and Knut Sørensen's study of a Norwegian hacker culture similarly, but in their case intentionally, undermines what earlier seemed certainties about the ways in which computing is masculine.

> One of the reasons we became interested in this group was the conclusions from a study of female computer science students. They used the hackers as a metaphor for all the things they did not like about computer science: the style of work, the infatuation with computers leading to neglect of normal non-study relations, and the concentration on problems with no obvious relation to the outside world. Thus, the hackers emerged as a possibly important example of an extremely masculine technological culture.[13]

Contrary to expectation, Håpnes and Sørensen encountered practices and attitudes which straddle gender lines: the hackers turned out to be competitive but communal and mutually supportive, "hard masters" yet open to the approaches characteristic of "soft mastery," fascinated by the machine yet concerned to write useful programs. The findings both confirm and confound the pictures painted by Joseph Weizenbaum and Sherry Turkle, touchstones for assays of this sort. "Compared to Turkle's description of American MIT hackers," they write, "the Norwegian hackers appear as less extreme and more 'feminine.' " But, they go on, that might have something to do with differences in the national cultures.[14]

Perhaps it does, but then we would have a curious separation of culture from its material base. Something is missing here. Although Håpnes and Sørensen refer at points to specific technical practices, they do not explore the role of the programming environment in enabling or even encouraging the behavior and attitudes they observed. They note that the hackers eschew Pascal and COBOL in favor of C and that they belittle the Macintosh while touting the virtues of their Sun workstations. But it is quite likely not the workstation alone that they are praising, because C does not stand by itself on that platform. It seems likely that the hackers are programming in the UNIX environment, which was quite consciously designed to foster certain patterns of behavior both individually and collectively. Since the turn of the 1980s, it has been the dominant operating and programming system in academic computing and, as a model at least, in the microcomputer software industry. Indeed, in the form of Linux it has recently begun to eat away at the near monopoly of Microsoft Windows.

Anyone familiar with UNIX will recognize it in the behavior of the Norwegian hackers. UNIX began as a system for sharing files and thus for communal work. Although Ken Thompson and Dennis Ritchie provided the basic system and the C language, respectively, the development of UNIX was a collaborative effort among a relatively large number of people both at Bell Labs and later at Berkeley. The source files were open to all, and anyone could decide to enhance or modify any part of the system. With that freedom came responsibility in accordance with the rule that whoever touched a routine assumed the task of maintaining it for the others until someone else decided to make a change. Membership in the community meant mutual support.[15]

The guiding metaphor of UNIX is the toolbox. UNIX is an environment for artisans, who craft software from a variety of small programs that "do one thing and do it well" but also can be linked to one another in sequence or "pipe," each taking input from its predecessor and supplying output to its successor. Many of these programs constitute "little languages" and emphasize the verbal aspects of computing. UNIX thus encourages what developers refer to as "rapid prototyping" but what Turkle and others would immediately recognize as "bricolage." Pipes and the little languages make programming an interactive enterprise, allowing the programmer to piece together solutions to problems before committing to a final program. That is what makes it appealing to Håpnes's and Sørensen's hackers, but it is also what ought to make it attractive to Estrin as a programming environment for women.

UNIX as a mainstream culture of computing seems to present considerable ambivalence here, and thus provides an interesting perspective from which to view, as a third example, Ulrike Erb's "Exploring the Excluded: A Feminist Approach to Opening New Perspectives in Computer Science." Erb undertook an empirical study of "[German] female computer scientists, who, in spite of the hacker culture and women's marginalization, have successfully made their way in computer science." She found that a "techno-centered" culture of computing acted as an obstacle to her subjects, who had sought access to the field by way of theoretical computer science:

> The interviews show that even among computer scientists it remains vague what it means to be competent in computer technology or to be a "technical insider": Does it mean sitting all day and night in front of the computer? Does it mean knowing every bit and byte or knowing the latest software? Or does it mean knowing how to construct blinking surfaces? In the interviews it became evident that in fact all this is associated with technical competence, while on the other hand knowledge on [sic] system design like requirements engineering, specification and design of algorithms is not thought to belong to technical competence.[16]

Erb emphasizes that this was often a matter of perception. In actual practice, the women displayed much of the technical competence they denied possessing, largely because they did not want to be associated with the work habits it connotes. By the same token, rapid prototyping did seem to attract more attention and praise from the profession at large than did more considered specification and design based on inquiry into users' needs and responses, an approach to which the women felt particularly drawn.

One would like to ask how they feel about UNIX. On the one hand, the UNIX culture rewards technical virtuosity and fosters rapid prototyping and exploratory approaches to design. On the other hand, it reflects fundamental theoretical concerns. Among those who contributed to it in both spirit and substance are Alfred Aho, John Hopcroft, and Jeffrey Ullman, who teamed in several various combinations to produce standard texts in computer science, including the theory of automata and formal languages, the design of algorithms, and the theory and design of compilers. Those texts can be found in the computer science section of German as well as American bookstores. Other contributors to UNIX have written equally accessible standard treatises in computational complexity. UNIX is theory-based in ways that make it a model for the approach to computing that Erb's women claim to admire and that they see as their means of access to the field. And it is both the operating system and the programming environment in which AT&T maintains its Electronic Switching System, the backbone of the U.S. telephone system, designed very much with the user in mind.

UNIX aside, Erb's point quoted above poses another problem for an analysis based on gender. The areas of requirements analysis, specification, and high-level design have formed part of software engineering from its inception in the early 1970s. Since the mid-1980s they have tended to dominate the attention of software engineers, both in attracting its leading practitioners and in forming the focus of what those practitioners consider best practice. No one doubts the crucial role of formal requirements analysis and specification; every textbook in software engineering emphasizes it, and the bulk of the research effort in the field has gone toward the development of tools to facilitate the work. In software engineering as a subdiscipline of computer science, formal methods of requirements analysis, specification, and design have drawn the major share of professional attention and recognition. As far as the theoretical community is concerned, Erb's women view the field in a way that would be difficult to corroborate by more or less objective evidence of its actual practice. So there is a problem of perception here, too.

That said, it is also the case that one of the dilemmas of software engineering from its beginnings is that what the theoretical community deems best practice is seldom reflected in the way that large software projects have actually been carried out. Studies have repeatedly shown that the most serious and most expensive errors occur in the early stages, in particular in setting down clearly what the system is supposed to do. Yet, they also show that only a small fraction of project time is devoted to those early stages. Most commonly, technical proficiency does indeed take the place of thoughtful analysis and design, and the outcome is often disastrous. But, in principle at least, the discipline deplores the problem. To that extent, Erb's subjects are in the mainstream, not bucking the current.

Software, Engineering, and Software Engineering

Erb concludes her article by urging that "there is a great need for women's research and in particular for women's research done by female computer scientists. In this way women's studies can help to reveal the excluded and to integrate the excluded in order to enrich computer science by means of the forgotten perspectives." Erb does not specify what perspectives she believes have been ignored or forgotten, and all three articles make it hard to see what they might be, since none of the usual suspects seems on close examination to be missing from the lineup. By contrast, the current state of software engineering suggests how her general call for reform might be made specific.[17]

The research of Lucy Suchman and her colleagues on participatory design is concerned, among other things, with situated cognition, or knowledge and skill that are invoked in specific circumstances. They are often tacit and hence go unnoticed or undervalued in studies of use and practice on which the design of computer systems is based. Those studies focus on what can be formulated in general terms: one does not usually computerize this or that traffic control room, but traffic control rooms in general. For historical and cultural reasons, women's work and their ways of carrying it out quite often disappear from view in this move from the situated to the general, e.g., the experience and judgment underlying clerical tasks, or the often critical decisions routinely made by nurses but not allocated to them professionally. This is where Erb sees promise in general. If we look at the professed interests of her subjects, that promise assumes a much more definite shape. We need to back up just a bit to see it.[18]

Between 1955 and 1970, through a pattern of interaction that remains largely unanalyzed, the use of computers spread widely while the computer industry expanded. The ever-wider variety of applications, combined with the increasing sophistication of computer systems themselves, put an exponentially growing pressure on programmers to keep pace with the demand. By the end of the 1960s, almost every large-scale program-ming project had fallen behind schedule, exceeded its budget, and failed to meet its specifications. Software was not keeping pace with hardware. There were never enough programmers, yet managers were learning that, as Fred Brooks put it in his famous law, "Adding manpower to a late soft-ware project makes it later."[19]

The dominant response in the industry was to view the situation as a problem in production, to be solved by some form of engineering. In 1968, that view started on the path to institutionalization with the con-vening of the first NATO Software Engineering Conference in Garmisch, Germany; a second followed in 1969 in Rome. "The phrase 'software engineering' was deliberately chosen as being provocative, in implying the need for software manufacture to be based on the types of theoreti-cal foundations and practical disciplines that are traditional in the es-tablished branches of engineering." Although individual uses of the term "software engineering" predated these conferences, they established it as a field of computing, and in the intervening thirty years it has acquired most of the hallmarks of a professional engineering discipline: societies (or groups within societies), journals, conferences, textbooks, curricu-lum, etc. One thing is missing, however, and it nags at practitioners: they cannot agree among themselves that it is in fact an engineering disci-pline and, if it is not, what would make it so. In the meantime, the prob-lems software engineering was supposed to solve have continued or grown even worse.[20]

One can take several different and revealing directions in exploring this situation. For present purposes, three observations may suffice. First, stat-ing the problem so as to make its solution a form of engineering commit-ted practitioners of the new field to historical and cultural models that have traditionally been associated with men. One model was engineering as applied science, which led to sustained efforts to reduce software develop-ment to mathematically based programming tools. Another model was in-dustrial engineering, which fostered the notion of the "software factory" along both Taylorist and Fordist lines. Closely allied to the industrial model was the view of engineering as project management, also with its Taylorist forms of organization of production by chain of command. Other models were open at the time, among them architecture and craft practice, but they received little attention.[21]

Second, no woman participated in either conference, although sev-eral women have subsequently established a strong presence in the field, precisely in the areas preferred by Erb's women, namely, requirements analysis, specification, and design. Interestingly, in her account they construe those areas as belonging to computer science, that is, as pertaining to theoretical analysis rather than technical experimentation. Theory, to repeat one of Erb's main findings, provides them with access to a world of computing that they otherwise view as uninviting or even hostile.[22]

That leads to the third and concluding point. The goal of software development is to model a portion of the real world on the computer. The process begins with the analysis and specification of just what one wants to model and how the model is to behave.*

That involves an understanding not of computers but of the real-world situation in question. It means in particular an understanding of what the people in that situation are trying to accomplish, what they contribute to accomplishing it, and how a computer system would help them to do it better or at least more efficiently. That is not what one learns in studying computer science; that is not what computer science is about.[23]

To accomplish that task requires that one know how to observe and to listen. To take a lead from Suchman, it requires being able to see what certain perspectives render invisible, to hear what certain discourses render inaudible. It places a premium on asking how the system is currently working before dictating how it should work in the future. Here, it seems to me, is where feminism could make a difference to one field of engineering (whether or not it is in fact engineering). Feminist analysis has brought out the ways in which a world built by men hides the ways in which women make it work, and it is the working world that computers must capture if they are going to enhance all our lives.

---

* Editorial note: Several paragraphs of text and a diagram have been removed here. They summarized material found elsewhere in this book in chapter 7, "Finding a History for Software Engineering" and chapter 13, "Software as Science—Science as Software."

PART THREE

The Structures of Computation

Computing and Mathematics at Princeton in the 1950s

Editorial note: This short paper was originally published in French, though the text here is Mahoney's original English version. Its second half is a brief summary of the central story presented in "Computer Science: The Search for a Mathematical Theory" and "Computers and Mathematics." However, I feel that this provides a useful and accessible introduction to the argument Mahoney makes in the longer papers as well as an interesting alternate framing of these developments as deeply rooted in Princeton's own history.

IN THE 1950S Princeton University was undergoing a transition from a predominantly liberal arts college to a research university. The faculty in science and engineering had distinguished themselves during the war and returned to Princeton eager to continue their well-funded investigations. The federal government, persuaded of the value of a scientific infrastructure, was equally eager to provide the funding. Over the next fifteen years, Princeton began the move to Big Science, as it became the home of an accelerator, a jet propulsion laboratory, a fusion reactor, and the professional staffs needed to support them. Government agencies paid for more than hardware. Even before the establishment of the National Science Foundation, contracts with the research offices of the armed services and with the Atomic Energy Commission included generous funding for faculty and their graduate assistants. Princeton was far from alone in enjoying the new prestige of the sciences and the funding that came with it. Its experience was being repeated across the country. But Princeton had a special advantage. Together with the Institute for Advanced Study, formally separate but closely allied in practice, it had a tradition of excellence in the mathematical sciences. It was among the first places to have a computer. And it was the only place that had John von Neumann.

Princeton's Computer

Von Neumann returned to Princeton following the Second World War with a variety of new projects in mind. In particular, he wanted to build a computer. His brief collaboration with John Mauchly and J. Presper Eckert, the creators of ENIAC at the University of Pennsylvania, had led to the famous "Draft Report on the EDVAC" (1945), in which von Neumann laid down the basic architecture that has since come to bear his name. He persuaded the Institute for Advanced Study to enter into a joint project with RCA's Sarnoff Laboratory and with Princeton University to translate his design into a working machine, which could then serve as the basis for research into the scientific possibilities of high-speed numerical computation, especially in the area of meteorology. Additional funding came from the army, the navy, and later the Atomic Energy Commission.[1]

First proposed in 1945, the IAS computer took six years to build, but even before completion it was being replicated at laboratories and universities across the U.S. and around the world, including Moscow's Academy of Sciences. Although the first calculations were done for the Los Alamos laboratories, von Neumann intended his machine for experiments in computation rather than production computing. Believing that "hydrodynamical problems are the prototype for anything involving non-linear partial differential equations," he focused at first on numerical meteorology, an area of obvious interest to his military sponsors. But he was interested in the range of problems that could be addressed and so made the computer available to researchers in numerical methods, statistics, traffic simulation, and even history of astronomy. His friend and Princeton colleague Eugene Wigner (Nobel Prize 1963) used the IAS computer for statistical studies of the wave functions of quantum mechanical systems. Martin Schwarzschild of the university's observatory explored the interiors of stars by means of numerical models. These were problems for which no analytical solution was known, and for von Neumann they were a test of the value of the computer to scientific research. Princeton's astrophysicists in particular took advantage of the new tool and the possibilities it presented for modeling phenomena unreachable by direct experiment. They would account for up to half the time on the IAS computer for the three years (1956–1959) it was operated by the university.

Despite the early advantage, neither the institute nor the university became a center of research in computing per se. At both institutions, the focus lay on the use of the computer as a tool for science rather than as an object of study in its own right. No more computers were designed and built in Princeton. When the IAS computer was retired at the end of the 1950s, the university turned to IBM for a new system to support the work of its scientists and engineers. Yet, many of the new directions in computer science that were pursued elsewhere have a Princeton connection made in the 1950s. Many of the people who shaped the computer as a scientific instrument and as a subject of scientific inquiry studied or taught at Princeton during the period and later recalled the inspiration and encouragement they had encountered here.

Among the several new areas of research that von Neumann pursued in the decade after the war, three had special importance both for computing and for Princeton: computation of non-linear problems, game theory, and the theory of automata. The first was built into the meteorology project and other applications of the IAS computer. His game theory dated originally from a seminal paper in 1928 but had languished until 1940 when conversations with Oskar Morgenstern, a Vienna-trained member of Princeton's Economics Department, stimulated von Neumann's interest in the application of game theory to economics. The result of their collaboration over the next several years was the classic *Theory of Games and Economic Behavior*. At first, economists in general and especially at Princeton showed little interest in the work. But in the late 1940s it engaged the enthusiasm of several faculty and graduate students in the mathematics department, who explored the relation of game theory to linear and non-linear programming and gave the new field its definitive shape during the 1950s. One of those students, John Nash, won the 1994 Nobel Prize in Economics for work done then.[2]

New Directions in Mathematics

What was striking about the Princeton mathematics department in the 1950s was its openness to new directions in the field and to applications of mathematics to new areas, indeed to applied mathematics in general. It was symbolized perhaps by the department's two chairmen during the period, Solomon Lefschetz and his student and successor, Albert W. Tucker. Both had established their careers as topologists but were now turning in new directions. In 1946 Lefschetz initiated a long-term project, funded by the Office of Naval Research (ONR), on non-linear differential equations with particular focus on non-linear oscillations, the theory of operators, and combinatorial problems in logistics. A few years later, Tucker started the Princeton Logistics Project, also under ONR sponsorship, with a focus on the geometry and computations of solutions of two-person and n-person games, network theory, and variations and perturbations of "linear programming" methods. Working in close collaboration, their two projects brought a variety of mathematical fields to bear on new approaches to non-linear systems. While maintaining the department's traditional strength in analysis and topology, they fostered wide-ranging explorations into dynamical systems, mathematical programming, decision theory, and computation. Several of the people involved went on to the RAND Corporation, where they continued their work in the area, often in collaboration with the Princeton groups.

It was a time of new possibilities, made all the more exciting by the openness of the Princeton mathematical community to new ideas and approaches. Martin Shubik, one of the seminal figures in computer-based experimental economics, came to Princeton in 1949 to study game theory as a graduate student in economics but found little interest in his own department. The mathematics department became his intellectual home. As he later reminisced,

> The general attitude around Fine Hall [the home of the department] was that no one really cared who you were or what part of mathematics you worked on as long as you could find some senior member of the faculty and make a case to him that it was interesting and that you did it well. . . . To me the striking thing at that time was not that the mathematics department welcomed game theory with open arms—but that it was open to new ideas and new talent from any source, and it could convey a sense of challenge and a belief that much new and worthwhile was happening.[3]

Perhaps the most extreme example was Marvin Minsky's dissertation in 1954 on the "Theory of Neural-Analog Reinforcement Systems," in which he began his explorations into what would soon become artificial intelligence. Tucker agreed to serve as adviser. Minsky was going where mathematicians had not gone, but as Tucker later recalled, "he had these ideas, and I for one felt that it was more useful to the world to have him develop these ideas, which were completely original, than to do something, say, in topology, which he could very easily have done."[4]

In 1956, Minsky teamed up with another Princeton Ph.D., John McCarthy, and with Claude Shannon of Bell Laboratories and Nathaniel Rochester of IBM, to propose a summer institute at Dartmouth on a subject they called "artificial intelligence." McCarthy, who had done a dissertation on differential equations under Lefschetz but who had also been active with the game theorists, subsequently became interested in formal reasoning on the computer. In a related effort to enlist support in the area, he had joined with Shannon to edit a volume on *Automata Studies*, which was supported in part by the Logistics Project and which appeared in Princeton's *Annals of Mathematics Studies*. Although the contributions to the volume did not take the direction McCarthy had hoped, they did lay a foundation for the newly emerging theory of computation. Again, the seminal inspiration came from von Neumann.[5]

The Science of Computers

In designing the IAS computer, von Neumann worked from an overall vision that far exceeded current technical capabilities. Beginning in the late 1940s and continuing until his death in 1957, he explored the conditions under which computers might detect and correct their own failures, might replicate themselves, and might even evolve to ever-more complex forms. The model, of course, was the living organism and in particular the human brain. While von Neumann pictured computers assembling copies of themselves in a sea of physical parts, his friend Stanislaw Ulam suggested the model of a cellular automaton, a multidimensional array of finite machines changing state in response to the states of their immediate neighbors. Although the University of Michigan would be home to research in that area for the next twenty years, von Neumann's discussion of "growing automata" refocused attention on finite automata, and there Princeton mathematicians had fundamental things to say.

From the beginning, von Neumann insisted that the computer not only met the computational needs of traditional mathematical science and provided new means of handling previously intractable mathematical systems, but also opened areas of investigation previously thought inaccessible to mathematics, for example, the growth of organisms and the workings of the human mind. Here, the computer would serve as a model of these systems, and an understanding of the model would rest on an adequate theory of the computer itself. But that, too, was largely uncharted territory, where von Neumann himself felt lost. "We are," he proclaimed, "very far from possessing a theory of automata which deserves that name, that is, a properly mathematical-logical theory."

What would such a theory look like? At first glance it would seem to take mathematics onto unfamiliar and difficult terrain.

> There exists today a very elaborate system of formal logic, and, specifically, of logic as applied to mathematics. This is a discipline with many good sides, but also with certain serious weaknesses. This is not the occasion to enlarge upon the good sides, which I certainly have no intention to belittle. About the inadequacies, however, this may be said: Everybody who has worked in formal logic will confirm that it is one of the technically most refractory parts of mathematics. The reason for this is that it deals with rigid, all-or-none concepts, and has very little contact with the continuous concept of the

real or of the complex number, that is, with mathematical analysis. Yet analysis is the technically most successful and best-elaborated part of mathematics. Thus formal logic is, by the nature of its approach, cut off from the best cultivated portions of mathematics, and forced onto the most difficult part of the mathematical terrain, into combinatorics.

The theory of automata, of the digital, all-or-none type, as discussed up to now, is certainly a chapter in formal logic. It will have to be, from the mathematical point of view, combinatory rather than analytical.[6]

Von Neumann did not elaborate on the nature of that combinatory mathematics, nor suggest from what areas of current mathematical research it might be drawn.

As noted above, the Princeton mathematics department was already engaged in forging new relationships between analysis and combinatorics. In Alonzo Church, it had a leading figure in mathematical logic, whose students of two generations played leading roles in constructing the mathematical theory von Neumann was seeking. While still graduate students employed during the summer at IBM in 1957, Michael Rabin and Dana Scott wrote their seminal paper on "Finite Automata and Their Decision Problems."

Princeton during the 1950s fits into the story of "science and the computer" in a complex and indirect way, perhaps best illustrated by the story of the lambda calculus. It was created by Church during the two years that he was away from Princeton between his graduate studies and his career on the faculty. He conceived it as a means of eliminating free variables from logical expressions and hence of making it possible to denote predicates without reference to the nature of their variables. In his early career at Princeton, the lambda calculus formed the basis of his logic and an area of active research by his students. One of these, Stephen C. Kleene, showed in his doctoral dissertation how one could do arithmetic in the system. Church believed at first that the lambda calculus could avoid Russell's paradox and then Gödel's dilemma, but Church's students soon established that it did neither, and in 1937 Alan Turing, who had come to Princeton to study with Church, showed that the lambda calculus was computationally equivalent to the Turing machine. By the late 1940s, Church had relegated the lambda calculus to a footnote in his *Introduction to Mathematical Logic*, and it appeared only in special cases in Kleene's own *Introduction to Metamathematics*.[7]

In 1958, John McCarthy was devising a new computer language for his research into mechanical theorem proving and what he called his "advice taker," a program for symbolic reasoning. Needing a notation for applying an abstract function to an indefinitely long list of arguments, he turned to the list of *Annals of Mathematics Studies* and there found Church's treatise on the lambda calculus. By McCarthy's own account, he did not understand much of the theory itself, but he did adopt the notation as a metalanguage for his new language, LISP, which quickly became the language of choice for artificial intelligence programming and established the paradigm of functional programming.

But in addition to artificial intelligence, McCarthy was also interested in a mathematical theory of computation which would take account of the semantics of programs, that is, of processes or functions by which programs transformed input into output. While he thought in terms of LISP, others attracted to his agenda of formal semantics focused on the lambda calculus itself. The problem with the lambda calculus as a theoretical foundation, as commentators on McCarthy's work had pointed out, was that it did not have a mathematical model. In and of itself, it did not constitute a mathematical theory of computation, as McCarthy had claimed.

In the late 1960s, the problem came under the scrutiny of Dana Scott, who had turned again to questions of computation following work in logic and model theory. Critical of the efforts of Christopher Strachey in Cambridge and Jaco de Bakker in Amsterdam to base a mathematical semantics on the lambda calculus, Scott set out to show that it would not work, only then to discover the mathematical model that lambda calculus had lacked. The resulting theory of denotational semantics based on continuous lattices forged new links between the theory of computation and modern abstract algebra, in particular category theory. It also confirmed the lambda calculus as the foundation of functional programming languages. None of this happened in Princeton in the 1950s, but it began there.*

* Editorial Note: The two final paragraphs of the original article are omitted here.

CHAPTER TEN

Computer Science: The Search for a Mathematical Theory

THE DISCIPLINE OF COMPUTER SCIENCE has evolved dynamically since the creation of the device that now generally goes by that name, that is, the electronic digital computer with random-access memory in which instructions and data are stored together and hence the instructions can change as the program is running. The first working computers were the creation not of scientific theory, but of electrical and electronic engineering practice applied to a long-standing effort to mechanize calculation. The first applications of the computer to scientific calculation and electronic data processing were tailored to the machines on which they ran, and in many respects they reflected earlier forms of calculating and tabulating devices. Originally designed as a tool of numerical calculation, in particular for the solution of non-linear differential equations for which no analytical solution was available, the computer led immediately to the new field of numerical analysis. There, speed and efficiency of computation determined the initial agenda, as mathematicians wrestled large calculations into machines with small memories and with basic operations measured in milliseconds.

As the computer left the laboratory in the mid-1950s and entered both the defense industry and the business world as a tool for data processing, for real-time command and control systems, and for operations research, practitioners encountered new problems of non-numerical computation posed by the need to search and sort large bodies of data, to make efficient use of limited (and expensive) computing resources by distributing tasks over several processors, and to automate the work of programmers who, despite rapid growth in numbers, were falling behind the even more quickly growing demand for systems and application software. The emergence during the 1960s of time-sharing operating systems, of computer graphics, of communications between computers, and of artificial intelligence increasingly refocused attention from the physical machine to abstract models of computation as a dynamic process.

Most practitioners viewed those models as mathematical in nature and hence computer science as a mathematical discipline. But it was mathematics with a difference. While insisting that computer science deals with the structures and transformations of information analyzed mathematically, the first Curriculum Committee on Computer Science of the Association for Computing Machinery (ACM) in 1965 emphasized the computer scientists' concern with effective procedures:

> The computer scientist is interested in discovering the pragmatic means by which information can be transformed to model and analyze the information transformations in the real world. The pragmatic aspect of this interest leads to inquiry into effective ways to accomplish these at reasonable cost.[1]

A report on the state of the field in 1980 reiterated both the comparison with mathematics and the distinction from it:

> Mathematics deals with theorems, infinite processes, and static relationships, while computer science emphasizes algorithms, finitary constructions, and dynamic relationships. If accepted, the frequently quoted mathematical aphorism, "the system is finite, therefore trivial," dismisses much of computer science.[2]

Computer people knew from experience that "finite" does not mean "feasible" and hence that the study of algorithms required its own body of principles and techniques, leading in the mid-1960s to the new field of computational complexity. Talk of costs, traditionally associated with engineering rather than science, involved more than money. The currency was time and space, as practitioners strove to identify and contain the exponential demand on both as even seemingly simple algorithms were applied to ever-larger bodies of data. Yet, central as algorithms were to computer science, the report continued, they did not exhaust the field, "since there are important organizational, policy, and nondeterministic aspects of computing that do not fit the algorithmic mold."

There have been some notable exceptions to the notion of computing as essentially a mathematical science. In a widely read letter to *Science* in 1967, Allan Newell, A. J. Perlis, and Herbert Simon, each a recipient of the ACM's highest honor, the Turing Award, argued that the computers were as much phenomena as instruments and hence that the study of them constituted an empirical science. Other winners have echoed that view, among them Marvin Minsky, one of the creators of artificial intelligence, and Donald Knuth, whose two-volume *Art of Computer Programming* has long been the standard reference tool for programmers.[3]

Thus, in striving toward theoretical autonomy, computer science has always maintained contact with practical applications, blurring commonly made distinctions among science, engineering, and craft practice. That characteristic makes the field both a resource for the reexamination of these distinctions and an elusive subject to encompass in a short historical account. What follows, therefore, focuses on the core of the search for a theory of computing, namely the effort to express the computer and computation in mathematical terms adequate to practical experience and applicable to it.

In tracing the emergence of a discipline, it is useful to think in terms of its *agenda*, that is, what practitioners of the discipline agree ought to be done, a consensus concerning the problems of the field, their order of importance or priority, the means of solving them, and perhaps most importantly, what constitute solutions. Becoming a recognized practitioner of a discipline means learning the agenda and then helping to carry it out. Knowing what questions to ask is the mark of a full-fledged practitioner, as is the capacity to distinguish between trivial and profound problems. Whatever specific meaning may attach to "profound," generally it means moving the agenda forward. One acquires standing in the field by solving the problems with high priority, and especially by doing so in a way that extends or reshapes the agenda, or by posing profound problems. The standing of the field may be measured by its capacity to set its own agenda. New disciplines emerge by acquiring that autonomy. Conflicts within a discipline often come down to disagreements over the agenda: what are the really important problems? Irresolvable conflict may lead to new disciplines in the form of separate agendas.

As the shared Latin root indicates, agendas are about action: what is to be *done*? Since what practitioners do is all but indistinguishable from the way they go about doing it, it follows that the tools and techniques of a field embody its agenda. When those tools are employed outside the field, either by a practitioner or by an outsider borrowing them, they bring the agenda of the field with them. Using those tools to address another agenda means reshaping the latter to fit the tools, even if it may also lead to a redesign of the tools. What gets reshaped, and to what extent, depends on the relative strengths of the agendas of borrower and borrowed.

Machines That Compute

At the outset, computing had no agenda of its own from which a science might have emerged. As a physical device, the computer was not the product of a scientific theory from which it inherited an agenda. Rather, computers and computing posed a constellation of problems that intersected with the agendas of various fields. As practitioners of those fields took up the problems, applying to them the tools and techniques familiar to them, they gradually defined an agenda for computer science. Or, rather, they defined a variety of agendas, some mutually supportive, some orthogonal to one another, and some even at cross-purposes. Theories are about questions, and where the nascent subject of computing could not supply the next question, the agenda of the outside field often provided it.

When the computer came into existence, there were two areas of mathematics concerned with it as a subject in itself rather than as a means for doing mathematics. One was mathematical logic and the theory of computable functions. To make the notion of "computable" as clear and simple as possible, Alan Turing proposed in 1936 a mechanical model of what a human does when computing:

> We may compare a man in the process of computing a real number to a machine which is only capable of a finite number of conditions $q_1, q_2, \ldots,$ q_R which will be called "m-configurations." The machine is supplied with a "tape" (the analogue of paper) running through it, and divided into sections (called "squares") each capable of bearing a "symbol."[4]

Turing imagined, then, a tape divided into cells, each containing one of a finite number of symbols. The tape passes through a machine that can read the contents of a cell, write to it, and move the tape one cell in either direction. What the machine does depends on its current state, which includes a signal to read or write, a signal to move the tape right or left, and a shift to the next state. The number of states is finite, and the set of states corresponds to the computation. Since a state may be described in terms of five symbols (current state, input, output, shift right/left, next state), a computation may itself be expressed as a sequence of symbols, which can also be placed on the tape, thus making possible a universal machine that can read a computation and then carry it out by emulating the machine described by it.

Turing's machine, or rather his monograph, fitted into the current agenda of mathematical logic. The *Entscheidungsproblem* stemmed from David Hilbert's program of formalizing mathematics; as stated in the textbook he wrote with W. Ackermann:

The *Entscheidungsproblem* is solved when one knows a procedure by which one can decide in a finite number of operations whether a given logical expression is generally valid or is satisfiable. The solution of the *Entscheidungsproblem* is of fundamental importance for the theory of all fields, the theorems of which are at all capable of logical development from finitely many axioms.[5]

Having written the paper for W. H. A. Newman's senior course at Cambridge, Turing turned for graduate study to Alonzo Church at Princeton, who had recently introduced an equivalent notion of "computability," which he called "effective calculability." Turing subsequently showed that his machine had the same power as Church's lambda calculus or Stephen Kleene's recursive function theory for determining the range and limitations of axiom systems for mathematics.[6]

The other area of mathematics pertinent to the computer originated in a 1938 paper by Claude Shannon of MIT and Bell Telephone Laboratories.[7] In it Shannon had shown how Boolean algebra could be used to analyze and design switching circuits. It took some time for the new technique to take hold, and it came to general attention among electrical engineers only in the late 1940s and early 1950s. The literature grew rapidly in the following years, but much of it took a standard form. Addressed to an audience of electrical engineers, most articles included an introduction to Boolean algebra, emphasizing its affinity to ordinary algebra and the rules of idempotency and duality that set it apart. Either by way of motivation at the outset, or of application at the end, the articles generally included diagrams of circuits and of various forms of switches and relays, the former to establish the correspondence between the circuits and Boolean expressions, the latter to illustrate various means of embodying basic Boolean expressions in concrete electronic units. The articles gave examples of the transformation of Boolean expressions to equivalent forms, which corresponded to the analysis of circuits. Then they turned to synthesis by introducing the notion of a Boolean function and its expression in the normal or canonical form of a Boolean polynomial. The articles showed how the specification of a switching circuit in the form of a table of inputs and outputs can be translated into a polynomial, the terms of which consist of products expressing the various combinations of inputs. From the polynomial one could then calculate the number of switches required to realize the circuit and seek various transformations to reduce that number to a minimum.

As the notion of an algebraic theory of switching circuits took root, authors began shifting their attention to its inadequacies. They were not trivial, especially when applied to synthesis or design of circuits, rather than analysis of given circuits. For circuits of any complexity, representation in Boolean algebra presented large, complicated expressions for reduction to minimal or canonical form, but the algebra lacked any systematic, i.e., algorithmic, procedure for carrying out that reduction. The symbolic tool meant to make circuits more tractable itself became intractable when the number of inputs grew to any appreciable size.

Neither Turing's nor Shannon's mathematics quite fit the actual computer. Turing effectively showed what a computer could not do, even if given infinite time and space to do it. But Turing's model gave little insight into what a finite machine could do. As Michael Rabin and Dana Scott observed in their seminal paper on finite automata:

> Turing machines are widely considered to be the abstract prototype of digital computers; workers in the field, however, have felt more and more that the notion of a Turing machine is too general to serve as an accurate model of actual computers. It is well known that even for simple calculations it is impossible to give an *a priori* upper bound on the amount of tape a Turing machine will need for any given computation. It is precisely this feature that renders Turing's concept unrealistic.[8]

By contrast, Boolean algebra described how the individual steps of a calculation were carried out, but it said nothing about the nature of the computational tasks that could be accomplished by switching circuits. A theory of computation apposite to the electronic digital stored-program computer with random-access memory lay somewhere between the Turing machine and the switching circuit. Computer science arose out of the effort to fill that gap between the infinite and the finite, a region that had hitherto attracted little interest from mathematicians.

Just what and who would fill it was not self-evident. In a lecture "On a Logical and General Theory of Automata" in 1948, John von Neumann pointed out that automata and hence computing machines had been the province of logicians rather than mathematicians, and a methodological gulf separated the two:

> There exists today a very elaborate system of formal logic, and, specifically, of logic as applied to mathematics. This is a discipline with many good sides, but also with certain serious weaknesses. This is not the occasion to enlarge upon the good sides, which I certainly have no intention to belittle. About the inadequacies, however, this may be said: Everybody who has worked in formal logic will confirm that it is one of the technically most refractory parts of mathematics. The reason for this is that it deals with rigid, all-or-none concepts, and has very little contact with the continuous concept of the real or of the complex number, that is, with mathematical analysis. Yet analysis is the technically most successful and best-elaborated part of

mathematics. Thus formal logic is, by the nature of its approach, cut off from the best-cultivated portions of mathematics, and forced onto the most difficult part of the mathematical terrain, into combinatorics.

The theory of automata, of the digital, all-or-none type, as discussed up to now, is certainly a chapter in formal logic. It will have to be, from the mathematical point of view, combinatory rather than analytical.[9]

Von Neumann subsequently made it clear he wanted to pull the theory back toward the realm of analysis, and he did not expand upon the nature of the combinatory mathematics that might be applicable to it.

It was not clear what tools to use in large part because it was not clear what job to do. What should a theory of computing be about? Different answers to that question led to different problems calling for different methods. Or rather, it opened the theory of computing to different approaches, shaped as much by the taste and training of the theorist as by the parameters of the problem. As figure 7 illustrates, people came to the theory of computing from a variety of disciplinary backgrounds, most of them mathematical in form yet diverse in the particular kind of mathematics they used. Some of them were centrally concerned with computing; others took problems in computing as targets of opportunity for methods devised for other purposes altogether. In the long run, it all seemed to work out so well as to lie in the nature of the subject. But computers had no nature until theorists created one for them, and different theorists had different ideas about the nature of computers.

Theory of Automata

As von Neumann's statement makes clear, automata traditionally lay in the province of logicians, who had contemplated the mechanization of reasoning at least since Leibniz expounded the resolution of disagreements through calculation. Since logical reasoning was associated with rational thought, automata formed common ground for logicians and neurophysiologists. Before Turing himself made the case for a thinking machine and proposed his famous test in 1950, his original paper provoked Warren McCulloch and Walter Pitts in 1943 to model the behavior of connected neurons in the symbolic calculus of propositions and to show that in principle such "nervous nets" were capable of carrying out any task for which complete and unambiguous instructions could be specified. The behavior of the nets did not depend on the material of the components but only on their behavior and connectivity, and hence they could equally well be realized by switching circuits.[10]

Automata thus brought convergence to the agendas of the neurophysiologists and of the electrical engineers. Shannon's use of Boolean algebra to analyze and design switching circuits spread quickly in the late 1940s and early 1950s. In its original form, the technique applied to combinational circuits, the output of which depends only on the input. Placing a secondary relay in a circuit made its behavior dependent on its internal state as well, which could be viewed as a form of memory. Such sequential circuits could also be represented by Boolean expressions. For both circuits, the basic problem remained the reduction of complex Boolean expressions to their simplest form, which in some cases (but not all) meant the least number of components, a major concern when switches were built from expensive and unreliable vacuum tubes.

The analysis of sequential circuits gave rise then to the notion of a sequential machine, described in terms of a set of states, a set of input symbols, a set of output symbols, and a pair of functions mapping input and current state to next state and output, respectively. Starting in an initial state, a sequential machine read a sequence of symbols and outputted a string of symbols in response to it. The output was not essential to the model; one could restrict one's attention to the final state of the machine after reading the input. "Black-boxing" the machine shifted attention from the circuits to the input and output, and thus to its description in terms of the inputs it recognizes or the transformations it carries out on them. With the shift came two new sets of questions: first, what can one say about the states of a machine on the basis simply of its input-output behavior and, second, what sorts of sequences can a machine with a finite set of states recognize and what sorts of transformations can it carry out on them?

In 1951, working from the paper by McCulloch and Pitts and stimulated by the page proofs of von Neumann's lecture quoted above, Stephen C. Kleene established a fundamental property of finite-state automata, namely that the sets of tapes recognized by them form a Boolean algebra. That is, if L_1 and L_2 are sets of strings recognized by a finite automaton, then so too are their union, product, complements, and iterate or closure. Still working in the context of nerve nets, Kleene called such sets "regular events," but they soon came to be called "regular languages," marking thereby a new area of convergence with coding theory and linguistics.[11]

By the end of the 1950s a common mathematical model of the finite automaton had emerged among the convergent agendas, set forth in the seminal paper by Michael Rabin and Dana Scott mentioned above. Their formalism, aimed at retaining the flavor of the machine and its ties to

Turing machines, became the standard for the field, beginning, significantly, with tapes formed by finite sequences of symbols from a finite alphabet Σ. The class of all such tapes, including the empty tape Λ, is denoted T. A finite automaton A over an alphabet Σ is a quadruple (S, M, s_0, F), where S is a finite set of states, M is a function which for every pair consisting of a state and a symbol specifies a resulting state, s_0 is an initial state, and F is a subset of S comprising the final states. As defined, the automaton works sequentially on symbols, but by an obvious extension of M to sequences, it becomes a tape machine, so that $M(s,x)$ is the state at which the machine arrives after starting in state s and traversing the tape x. In particular, if $M(s_0,x)$ is one of the final states in F, the automaton "accepts" x. Let $T(A)$ be the set of tapes in T accepted, or defined, by automaton A, and let T be the class of all sets of tapes defined by some automaton.

Rabin and Scott pursued two related sets of questions: first, what can be determined about the set of tapes $T(A)$ accepted by a given automaton A, and, second, what is the mathematical structure of $T(A)$ and of the larger class T? Taking a lead from Kleene's results, they established that T constitutes a Boolean algebra of sets and hence that it contains all finite sets of tapes. Two central decision problems were now immediately solvable. First, if an automaton A accepts any tapes, it accepts a tape of length less than the number of its internal states, and therefore an effective procedure exists for determining whether $T(A)$ is empty (the "emptiness problem"). Second, if A is an automaton with r internal states, then $T(A)$ is infinite if and only if it contains a tape of length greater than or equal to r and less than or equal to $2r$; again, an effective procedure follows for determining whether the set of definable tapes is infinite.

At two points in their study, Rabin and Scott identified the set T with the semigroup, an algebraic structure that had until then received very little attention from mathematicians. But they did not make much use of the concept, approaching their model combinatorially and in essence constructing automata to meet the conditions of the problem. That reflected the agenda of mathematical logic, a subject still largely on the periphery of mathematics, outside its increasingly algebraic agenda. Yet, as the reference to the semigroup suggests, it was making contact with algebra. The link had been forged by way of coding theory.

Coding theory arose from another path-breaking paper of Shannon's, his "Mathematical Theory of Communication," first published in 1948. There Shannon offered a mathematical model of information in terms of sequences of signals transmitted over a line and posed the fundamental question of how to preserve the integrity of the sequence in the presence of distortion. The answer to that question took the form of sequential codes that by their very structure revealed the presence of errors and enabled their correction.

In a vaguely related way, circuit theory and coding theory verged on a common model from opposite directions. The former viewed a sequential machine as shifting from one internal state to another in response to input; the transition might, but need not involve output, and the final state could be used as a criterion of acceptance of the input. The main question was how to design the optimal circuit for given inputs, a matter of some importance when one worried about the cost and reliability of gates. Coding theory focused on the problem of mapping the set of messages to be transmitted into the set of signals available for transmission in such a way that the latter permit unambiguous retranslation when received. That means, first and foremost, that the stream of signals can be uniquely scanned into the constituents of the message. In addition, the process must take place as it occurs, and it should be immune to noise. That is, the code must be unambiguously decipherable as encountered sequentially, even if garbled in transmission.

It was in the context of coding theory that the agendas grouped around the finite automaton intersected with an important mathematical agenda centered in Paris. In "Une théorie algébrique du codage," Marcel P. Schützenberger, a member of Pierre Dubreil's seminar on algebra and number theory at the Faculté des Sciences in Paris, linked coding to the "fundamental algebraic structure" of Bourbaki:

> It is particularly noteworthy that the fundamental concepts of semigroup theory, introduced by P. Dubreil in 1941 and studied by him and his school from an abstract point of view, should have immediate and important interpretations on the level of the concrete realization of coding machines and transducers.[12]

Schützenberger translated the problem of coding into the conceptual structure of abstract semigroup theory by viewing a sequence of symbols as a semigroup and codes as homomorphisms between semigroups. The details are too complex to follow here, but the trick lay in determining that the structural properties of semigroups and their homomorphisms gave insight into the structure of codes. Schützenberger showed that they did and extended his analysis from the specific question of codes to the more general problem of identifying the structures of sequences of symbols with the structure of devices that recognize those structures, that is, to the problem of automata.

The resulting semigroup theory of machines thus gave a mathematical account of sequential machines that pulled together a range of results and perspectives. More importantly, perhaps, it suggested mathematical structures for which there was at the time no correlate in automata theory. Within Schützenberger's mathematical agenda, one moved from basic structures to their extension by means of other structures. In "Un problème de la théorie des automates" he set out the semigroup model of the finite automaton and then looked beyond it, aiming at an algebraic characterization based on viewing a regular language as a formal sum, or formal power series, Σx of the strings it comprises and considering the series as a "rational" function of the $x \in X$:

> We then generalize that property by showing that if, instead of being finite, S is the direct product of a finite set by $\mathbf{Z}$ (the additive group of integers) and if the mapping $(S, X) \rightarrow S'$ and the subset S_1 are defined in a suitable fashion, then the corresponding sum is in a certain sense "algebraic."[13]

That is, the extended sum takes the form $\Sigma<z,x>x$, where $<z,x>$ is the integer mapped to the string x. The terminology reflected the underlying model. Were the semigroup of tapes commutative, the sums would correspond to ordinary power series, to which the terms "rational" and "algebraic" would directly apply. The generalization was aimed at embedding automata in the larger context of the algebra of rings.

Initially, Schützenberger suggested no interpretation for the extension; that is, it corresponded to no new class of automata or of languages accepted by them. In fact, both the languages and the automata were waiting in the wings. They came on stage through the convergence of finite automata with yet another previously independent line of investigation, namely, the new mathematical linguistics under development by Noam Chomsky. The convergence of agendas took the concrete form of Schützenberger's collaboration with Chomsky on context-free grammars, then just getting under way. The result was the mathematical theory of formal languages.

Formal Languages

Formal language theory similarly rested at the start on an empirical base. Chomsky's seminal article, "Three Models of Language" (1956), considered grammar as a mapping of finite strings of symbols into all and only the grammatical sentences of the language:

> The grammar of a language can be viewed as a theory of the structure of this language. Any scientific theory is based on a certain finite set of observations and, by establishing general laws stated in terms of certain hypothetical constructs, it attempts to account for these observations, to show

how they are interrelated, and to predict an indefinite number of new phenomena. A mathematical theory has the additional property that predictions follow rigorously from the body of theory.[14]

The theory he was seeking concerned the fact of grammars, rather than grammars themselves. Native speakers of a language extract its grammar from a finite number of experienced utterances and use it to construct new sentences, all of them grammatical, while readily rejecting ungrammatical sequences. Chomsky was looking for a metatheory to explain how to accomplish this feat, or at least to reveal what it entailed. (Ultimately the physical system being explained was the human brain, the material basis of both language and thought.) In 1958 he made the connection with finite automata through a collaborative investigation with George Miller on finite state languages and subsequently joined forces with Schützenberger, then a visitor at MIT.

As powerful as the semigroup seemed to be in capturing the structure of a finite automaton, the model was too limited by its generality to handle the questions Chomsky was asking. It recognized languages produced by translations from tokens to strings of the form $\alpha \to x$, $\alpha \to x\beta$, (α, β non-terminals, x a string of terminals), by then called "finite state," or "regular" languages, but not languages generated by recursive productions such as $\alpha \to u\alpha v$, $\alpha \to x$, by which sequences may be embedded in other sequences an arbitrary number of times. These were the languages determined by the phrase structure grammars introduced by Chomsky in 1959, foremost among them the context-free grammars, and modeling them mathematically required more structure than the monoid could provide. That was reflected in the need for internal memory, represented by the addition of a second (two-way) tape. It was also reflected in the fact, quickly ascertained, that the languages are not closed under complementation or intersection and hence cannot be represented by something so tidy as a Boolean algebra.

What made the context-free grammars so important was the demonstration by Seymour Ginsburg and H. G. Rice in 1961 that large portions of the grammar of the new internationally designed programming language, ALGOL, as specified in John Backus's new formal notation, BNF (Backus Normal Form, or later Backus-Naur Form), based on the productions of Emil Post, are context-free. When the productions are viewed as set equations, the iterated solutions form lattices, which have as fixpoints the final solutions that correspond to the portions of the language specified by the productions. Of course, were ALGOL as a whole context-free, then the language so determined would consist of all syntactically valid programs. Here, Schützenberger could bring his model of formal power series to bear. In a group of papers in the early 1960s, he showed that the series he had characterized as "algebraic" were the fixpoint solutions of context-free grammars. The approach through formal power series also enabled him to show that the grammars were undecidably ambiguous and that they could be partitioned into two distinct subfamilies, only one of which fitted Chomsky's hierarchy.

Several agendas again converged at this point, as a mix of theoretical and practical work in both the United States and Germany on parenthesis-free notation, sequential formula translation for compilers, management of recursive procedures, and syntactic analysis for machine translation more or less independently led to the notion of the "stack" (in German, *Keller*), or list of elements accessible at one end only on a last in–first out (LIFO) basis. In the parlance of automata theory, a stack provided a finite automaton with additional memory in the form of a tape that ran both ways but moved only one cell at a time and contained only data placed there by the automaton during operation. Schützenberger and Chomsky showed that such "pushdown automata" correspond precisely to context-free languages and embedded the correspondence in Schützenberger's algebraic model.

The development, hand in hand, of automata theory and formal language theory during the late 1950s and early 1960s gave an empirical aspect to the construction of a mathematical theory of computation. The matching of Chomsky's hierarchy of phrase-structure grammars with distinct classes of automata enlisted such immediate assent precisely because it encompassed a variety of independent and seemingly disparate agendas, both theoretical and practical, ranging from machine translation to the specification of ALGOL and from noise-tolerant codes to natural language. These it united theoretically in much the same way that Newton's mechanics united not only terrestrial and celestial phenomena but also a variety of agendas in mathematics, mechanics, and natural philosophy. By the late 1960s, the theory of automata and formal languages had assumed canonical shape around an agenda of its own. In 1969, John Hopcroft and Jeffrey Ullman began their text, *Formal Languages and Their Relation to Automata*, by highlighting Chomsky's mathematical grammar, the context-free definition of ALGOL, syntax-directed compilation, and the concept of the compiler-compiler.

Since then a considerable flurry of activity has taken place, the results of which have related formal languages and automata theory to such an extent that it is impossible to treat the areas separately. By now, no serious study of computer science would be complete without a knowledge of the techniques and results from language and automata theory.

Far from merely complementing the study of computer science, the subject of automata and formal languages became the theoretical core of the curriculum during the 1970s, especially as it was embedded in such tools as lexical analyzers and parser generators. With those tools, what had once required years of effort by teams of programmers became an undergraduate term project.

While the bulk of that new agenda focused on grammars and their application to programming, part of it pulled the theory back toward the mathematics from which it had emerged. Increasingly pursued at a level of abstraction that tended to direct attention away from the physical system, the theory of automata verged at times on a machine theory of mathematics rather than a mathematical theory of the machine. "It appeared to me," wrote Samuel Eilenberg, co-creator of category theory, in the preface of his four-volume *Automata, Languages, and Machines*, "that the time is ripe to try and give the subject a coherent mathematical presentation that will bring out its intrinsic aesthetic qualities and bring to the surface many deep results which merit becoming part of mathematics, regardless of any external motivation." Characterized by constructive algorithms rather than proofs of existence, those results constituted, he went on, nothing less than a "new algebra, . . . contain[ing] methods and results that are deep and elegant" and that would someday be "regarded as a standard part of algebra." That is, its ties to the computer would disappear.[15]

Formal Semantics

Yet, even when the computer remained at the focus, some practitioners expressed doubt that formal language theory would suffice as a theory of computation, since computing involved more than the grammar of programming languages could encompass. As noted above, the practical activity of programming created the theoretical space to be filled by a mathematical model of computation. As the size and complexity of the programs expanded, driven by a widely shared sense of the all but unlimited power of the computer as a "thinking machine," so too did the expectations of what would constitute a suitable theory. One concept of a "mathematical theory of computation" stems from two papers in the early 1960s by John McCarthy, the creator of LISP and a progenitor of "artificial intelligence," yet another agenda of computing, with links both to neurophysiology and to logic. It was McCarthy who invoked Newton as historical precedent for the enterprise:

> In this paper I shall discuss the prospects for a mathematical science of computation. In a mathematical science, it is possible to deduce from the basic assumptions, the important properties of the entities treated by the science. Thus, from Newton's law of gravitation and his laws of motion, one can deduce that the planetary orbits obey Kepler's laws.[16]

In another version of the paper, he changed the precedent only slightly:

> It is reasonable to hope that the relationship between computation and mathematical logic will be as fruitful in the next century as that between analysis and physics in the last. The development of this relationship demands a concern for both applications and for mathematical elegance.[17]

The elegance proved easier to achieve than the applications. The trick was to determine just what constituted the entities of computer science.

McCarthy made clear what he expected from a suitable theory in terms of computing: first, a universal programming language along the lines of ALGOL but with richer data descriptions; second, a theory of the equivalence of computational processes, by which equivalence-preserving transformations would allow a choice of among various forms of an algorithm, adapted to particular circumstances; third, a form of symbolic representation of algorithms that could accommodate significant changes in behavior by simple changes in the symbolic expressions; fourth, a formal way of representing computers along with computation; and finally a quantitative theory of computation along the lines of Shannon's measure of information. What these criteria, recognizably drawn from the variety of agendas mentioned above, came down to in terms of McCarthy's precedents was something akin to traditional mathematical physics: a dynamical representation of a program such that, given the initial values of its parameters, one could determine its state at any later point. Essential to that model was the notion of converging on complexity by perturbations on a fundamental, simple solution, as in the case of a planetary orbit based initially on a two-body, central force configuration. Formal language theory shared with machine translation a similar assumption about converging on natural language through increasingly complex artificial languages.

For McCarthy, that program meant that automata theory would not suffice. Its focus on the structure of the tape belied both the architecture and the complexity of the machine processing the tape. As McCarthy put it:

Computer science must study the various ways elements of data spaces are represented in the memory of the computer and how procedures are represented by computer programs. From this point of view, most of the current work on automata theory is beside the point.[18]

It did not suffice to know that a program was syntactically correct. One should be able to give a mathematical account of its semantics, that is, of the meaning of the symbols constituting it in terms of the values assigned to the variables and of the way in which the procedures change those values.

Such an account required a shift in the traditional mathematical view of a function as a mapping of input values to output values, that is, as a set of ordered pairs of values. To study how procedures change values, one must be able to express and transform the structure of the function as an abstract entity in itself, a sequence of operations by which the output values are generated from the input values. Seeking an appropriate tool for that task, McCarthy reached back to mathematical logic in the 1930s to revive Alonzo Church's lambda calculus, originally devised in 1929 in an effort to find a type-free route around Russell's paradox. The attempt failed, and Church had long since abandoned the approach. The notation, however, seemed to fit McCarthy's purposes and, moreover, lay quite close structurally to LISP, which he had been using to explore mechanical theorem-proving.

The developments leading up to the creation of denotational, or mathematical, semantics in 1970 are too complicated to pursue in detail here. McCarthy's own efforts faltered on the lack of a mathematical model for the lambda calculus itself. It also failed to take account of the peculiar problem posed by the storage of program and data in a common memory. Because computers store programs and data in the same memory, programming languages allow unrestricted procedures which could have unrestricted procedures as values; in particular a procedure can be applied to itself. "To date," Dana Scott claimed in his "Outline of a Mathematical Theory of Computation" in 1970,

> no mathematical theory of functions has ever been able to supply conveniently such a free-wheeling notion of function except at the cost of being inconsistent. The main *mathematical* novelty of the present study is the creation of a proper mathematical theory of functions which accomplishes these aims (consistently!) and which can be used as the basis for the *metamathematical* project of providing the "correct" approach to semantics.[19]

One did not need unrestricted procedures to appreciate the problems posed by the self-application facilitated by the design of the computer. Consider the structure of computer memory, representing it mathematically as a mapping of contents to locations. That is, a state σ is a function mapping each element ℓ of the set L of locations to its value $\sigma(\ell)$ in V, the set of allowable values. A command effects a change of state; it is a function γ from the set of states S into S. Storing a command means that γ can take the form $\sigma(\ell)$, and hence $\sigma(\ell)(\sigma)$ should be well defined. Yet, as Scott insisted in his paper, "this is just an insignificant step away from the self-application problem $p(p)$ for 'unrestricted' procedures p, and it is just as hard to justify mathematically."

Recent work on interval arithmetic suggested to Scott that one might seek justification through a partial ordering of data types and their functions based on the notion of "approximation" or "informational content." With the addition of an undefined element as "worst approximation" or "containing no information," the data types formed a complete lattice, and monotonic functions of them preserved the lattice. They also preserved the limits of sequences of partially ordered data types and hence were continuous. Scott showed that the least upper bound of the lattice, considered as the limit of sequences, was therefore the least fixed point of the function and was determined by the fixed point operator of the lambda calculus. Hence self-applicative functions of the sort needed for computers had a consistent mathematical model. And so too, by the way, did the lambda calculus for the first time in its history. Computer science had come around full circle to the mathematical logic in which it had originated.

The Limits of Mathematical Computer Science

Beginning in 1975 and extending over the late 1970s, the Computer Science and Engineering Research Study, chaired by Bruce Arden of Princeton University, took stock of the field and its current directions of research and published the results under the title *What Can Be Automated?* The committee on theoretical computer science argued forcefully that a process of abstraction was necessary to understand the complex systems constructed on computers and that the abstraction "must rest on a mathematical basis" for three main reasons.

(1) Computers and programs are inherently mathematical objects. They manipulate formal symbols, and their input-output behavior can be described by mathematical functions. The notations we use to represent them strongly resemble the formal notations which are used throughout mathematics and systematically studied in mathematical logic.

(2) Programs often accept arbitrarily large amounts of input data; hence, they have a potentially unbounded number of possible inputs. Thus a program embraces, in finite terms, an infinite number of possible computations; and mathematics provides powerful tools for reasoning about infinite numbers of cases.

(3) Solving complex information-processing problems requires mathematical analysis. While some of this analysis is highly problem-dependent and belongs to specific application areas, some constructions and proof methods are broadly applicable, and thus become the subject of theoretical computer science.[20]

Defining theoretical computer science as "the field concerned with fundamental mathematical questions about computers, programs, algorithms, and information processing systems in general," the committee acknowledged that those questions tended to follow developments in technology and its application, and hence to aim at a moving target—strange behavior for mathematical objects.

Nonetheless, the committee could identify several broad issues of continuing concern, which were being addressed in the areas of computational complexity, data structures and search algorithms, language and automata theory, the logic of computer programming, and mathematical semantics. In each of these areas, it could point to substantial achievements in bringing some form of mathematics to bear on the central questions of computing. Yet, the summaries at the end of each section sounded a repeating chord. For all the depth of results in computational complexity, "the complexity of most computational tasks we are familiar with—such as sorting, multiplying integers or matrices, or finding shortest paths—is still unknown." Despite the close ties between mathematics and language theory, "by and large, the more mathematical aspects of language theory have not been applied in practice. Their greatest potential service is probably pedagogic, in codifying and given clear economical form to key ideas for handling formal languages." Efforts to bring mathematical rigor to programming quickly reach a level of complexity that makes the techniques of verification subject to the very concerns that prompted their development.

One might hope that the above ideas, suitably extended and incorporated into verification systems, would enable us to guarantee that programs are correct with absolute certainty. We are about to discuss certain theoretical and philosophical limitations that will prevent this goal from ever being reached. These limitations are inherent in the program verification process, and cannot be surmounted by any technical innovations.

Mathematical semantics could show "precisely why [a] nasty surprise can arise from a seemingly well-designed programming language," but not how to eliminate the problems from the outset. As a design tool, mathematical semantics was still far from the goal of correcting the anomalies that gave rise to errors in real programming languages.

If computers and programs were "inherently mathematical objects," the mathematics of the computers and programs of real practical concern had so far proved elusive. Although programming languages borrowed the trappings of mathematical symbolism, they did not translate readily into mathematical structures that captured the behavior of greatest concern to computing. To a large extent, theoretical computer science continued the line of inquiry that had led to the computer in the first place, namely, the establishment of the limits of what can be computed. While that had some value in directing research and development away from dead ends, it offered little help in resolving the problems of computing that lay well within those limits. For that reason, computer science remains an amalgam of mathematical theory, engineering practice, and craft skill.

CHAPTER ELEVEN

Extracts from *Computers and Mathematics: The Search for a Discipline of Computer Science*

Editorial note: This paper explains how a loosely coupled group of researchers, most notably Christopher Strachey, John McCarthy, and Dana S. Scott, turned lambda calculus into "a metalanguage for specifying and analyzing the semantics of programming language" and so went some way to providing computer science with its elusive mathematical underpinnings. Mahoney told essentially the same story in the "Formal Semantics" section of his later paper "The Structures of Computation and the Mathematical Structure of Nature." However, this earlier paper is more squarely focused on Strachey and McCarthy's work in the 1960s and includes a more detailed account of their contributions. I have eliminated a now redundant section at the end of this paper where Mahoney sketched issues covered more thoroughly in "The Structures of Computation" but retained the rest unchanged, including several quotes and explanatory passages very similar to those found in the later paper but impossible to remove without destroying the main thread of Mahoney's argument.

N A DISCUSSION on the last day of the second NATO Conference on Software Engineering held in Rome in October 1969, Christopher Strachey, director of the Programming Research Group at Oxford University, lamented that "one of the difficulties about computing science at the moment is that it can't demonstrate any of the things that it has in mind; it can't demonstrate to the software engineering people on a sufficiently large scale that what it is doing is of interest or importance to them." As example he cited the general ignorance or neglect by industry of the recursive methods that computer scientists took to be fundamental to programming. Blaming industry for failing to support research and faulting theorists for neglecting the real problems of practitioners, he went on to explore how the two sides might move closer together.[1]

Strachey's prescription is of less concern here than his diagnosis, which points to an interesting case study in the relation of science to technology in a field thought to be mathematical at heart. His remarks came at a significant point in the history of computing. It marked the end of two decades during which the computer and computing acquired their modern shape. As the title of a recent book on the early computer industry suggests, it was a time of *Creating the Computer*, when the question, "What is a computer, or what should it be?" had no clear-cut answer. By the late 1960s, the main points of that answer had emerged, determined as much in the marketplace as in the laboratory. At the same time, a consensus began to form concerning the nature of computer science, at least among those who believed the science should be mathematical. That consensus, reflected in the new category of "computer science" added to *Mathematical Reviews* in 1970, rested in part on theory and in part on experience, if not of the marketplace directly, then of actual machines applied to actual problems.[2]

As the computer was being created, then, so too was the mathematics of computing. When we think of the computer as a machine, it is not surprising that it should be an object of design, where available means and chosen purposes must converge on effective action. We are less accustomed to the idea that a mathematical subject might be a matter of design, that is, the matching of means to ends that themselves are open to choice. Yet, the agenda of the mathematical theory of computation changed as computers and programs grew in size and complexity during the first twenty years. If, as Saunders MacLane has said, mathematics is "finding the form to fit the problem," the mathematics of computing began with a search for the problem. Different people made different choices about what was significant, and the mathematics on which they drew varied accordingly. What follows is a first reconnaissance over this shifting terrain.

The electronic digital stored-program computer emerged from the convergence of two separate lines of development, each stretching back over several centuries but generally associated with the names of Charles Babbage and George Boole in the mid-nineteenth century. The first was concerned with mechanical calculation; the second involved mathematics and logic. In coming together, they brought two models of computation: the Boolean algebra of circuits created by Claude Shannon in 1938 and the mathematical logic of Turing machines devised by Alan Turing in 1936. In the early '50s, the new field of automata theory, inspired a decade earlier by the idea of the nervous system as a switching circuit and recently reinforced by the notion of the brain as computer, encompassed the two models at opposite ends of a spectrum ranging from finite deterministic machines to infinite or growing indeterministic machines.[3]

At the one end, beginning with the work of David A. Huffman, E. F. Moore, and G. H. Mealy, switching theory broadened its mathematical scope beyond Boolean algebra by gradually shifting attention from the internal structure of finite-state machines to the patterns of input they can recognize and thus to the notion of a machine as a mapping or partition of semigroups. By 1964, the field of algebraic machine theory was well established, with close links to the emerging fields that were reconstituting universal algebra. At the other end of the spectrum, during the mid-'60s Turing machines of various types became the generally accepted model for measuring the complexity of computations, a question that shifted attention from decidability to tractability and enabled a classification of problems in terms of the computing resources required for their solution. First broached by Michael O. Rabin in 1959 and '60, the subject emerged as a distinct field with the work of Juris Hartmanis and Richard E. Stearns in 1965 and acquired its full form with the work of Steven Cook and Richard Karp in the early '70s. The field has formed common ground for computer science and operations research, especially in the design and analysis of algorithms.[4]

As the two historical models of computation developed during the 1950s and '60s, they retained their distinctive characteristics. The one stayed close to the physical circuitry of computers, analyzing computation as it went on at the level of the switches. The other stood far away, considering what can be computed in the abstract, irrespective of the particular computer employed. By the mid-50s, however, a new—and, to some extent, unanticipated—complex of questions had arisen in the middle of these extremes. Programming was becoming an activity in its own right, prompting the development of programming languages and compiling techniques to ease the task of writing instructions for specific machines and to make programs transferable from one machine to another. The then dominant application to numerical analysis meant that most such languages would have a mathematical appearance, and the orientation to the programmer meant that they would use symbolic language. But some languages were aimed at insight into computing itself, and they emphasized the manipulation of symbols as opposed to numerical computation.

The explosion of languages over the decade 1955–1965, accompanied by the development of general techniques for their implementation and leading to programs of ever-greater size and complexity, established all these things as matters of practical fact. In doing so, they challenged computer scientists to give a mathematical account of them. The challenge grew increasingly urgent as problems of cost, reliability, and managerial control multiplied. The call for a discipline of "software engineering" in 1967 meant to some the reduction of programming to a field of applied mathematics.

At the same time, it was unclear what mathematics was to be applied and where. At first, automata theory seemed to hold great promise. In 1959, building on Stephen Kleene's general characterization of the events that prompt a response from the nerve nets specified by Warren McCulloch and Walter Pitts, Michael Rabin and Dana Scott showed that finite automata defined in the manner of Moore machines accepted the same "regular" sequences of characters (which corresponded to free semigroups) and extended the result to nondeterministic and other finite automata. Such regular expressions constituted the first of Noam Chomsky's hierarchy of phrase structure grammars, the fourth and highest of which corresponded to Turing machines. The suggested link between automata and mathematical linguistics, with its potential application to machine translation, sparked a burst of research. The early '60s saw the creation of pushdown and linearbounded automata to correspond to the intermediate levels of context-free and context-sensitive grammars. In 1965 formal language theory emerged as a field of computer science independent of the mathematical linguistics from which it had sprung. The new field provided a mathematical basis for lexical analysis and parsing of languages and thus gave theoretical confirmation to techniques such as John Backus' BNF, developed independently for specifying the syntax of ALGOL.[5]

Even as that work was going on, some writers began to argue that automata theory would not suffice as a mathematical theory of computation. In principle, the computer was a finite state machine; in practice it was an intractably large finite state machine. Moreover, it was not enough to know that a program is syntactically correct. A program is a function that maps input values to output values and hence is a mathematical object, the structure of which should itself be accessible to expression and analysis. Moreover, programs written in programming languages run on machines and must be translated by means of compilers into the languages of those machines. Both the functional structure of the program and its translation are a matter of semantics, a matter of what the statements of the program, and hence the program itself, mean. Three approaches to semantics emerged in the mid-'60s. The operational approach took the compiler itself to constitute a definition of the semantics of the language for a given machine, thus leaving open the question of establishing the equivalence of two compilers. The deductive approach, introduced by R. W. Floyd in 1967, linked logical statements to the steps of the program, thereby specifying its behavior as well as providing a means of verifying the program. Mathematical semantics aimed at a formal theory that would serve as a means of specification for compilers and as a metalanguage for talking about programs, algorithms and data.[6]

The proposal to make semantics the basis of a mathematical theory of computation came from two sources with different, though complementary emphases. John McCarthy was concerned with the structure of algorithms and how they might be compared with one another. Christopher Strachey spoke about the structure of computer memory and how programs alter its contents. Both men found common ground in a system of functional notation, the lambda calculus, first introduced by Alonzo Church in the early 1930s but subsequently abandoned by him when it did not fulfill his hopes of its serving as a foundation for mathematical logic. The use of the lambda calculus as a metalanguage for programs led to the first construction of a mathematical model for it, and it has subsequently come to be viewed as the "pure" programming language. None of this proceeded smoothly or directly, and it is worth looking at it in a bit more detail.[7]

At the Western Joint Computer Conference in May 1961, McCarthy proposed "A Basis for a Mathematical Theory of Computation." "Computation is sure to become one of the most important of the sciences," he began.

> This is because it is the science of how machines can be made to carry out intellectual processes. We know that any intellectual process that can be carried out mechanically can be performed by a general purpose digital computer. Moreover, the limitations on what we have been able to make computers do so far clearly come far more from our weakness as programmers than from the intrinsic limitations of the machines. We hope that these limitations can be greatly reduced by developing a mathematical science of computation.[8]

McCarthy made clear what he expected from a suitable theory: first, a universal programming language along the lines of ALGOL but with richer data descriptions; second, a theory of the equivalence of computational processes, by which equivalence-preserving transformations would allow a choice of among various forms of an algorithm, adapted to particular circumstances; third, a form of symbolic representation of algorithms that could accommodate significant changes in behavior by simple changes in the symbolic expressions; fourth, a formal way of representing computers along with computation; and finally a quantitative theory of computation along the lines of Shannon's measure of information.

McCarthy did not pretend to have met any of these goals, which spanned a broad range of currently separate areas of research. His work on the programming language LISP, however, had suggested a system of formalisms that allowed him to prove the equivalence of computations expressed in them. The formalisms offered means of describing functions computable in terms of base functions, using conditional expressions and recursive definitions. They included computable functionals (functions with functions as arguments), non-computable functions (quantified computable functions), ambiguous functions, and the definition both of new data spaces in terms of base spaces and of functions on those spaces, a feature that ALGOL, then the most theoretically oriented language, lacked. The system constituted the first part of McCarthy's paper; the second part set out some of its mathematical properties, a method called "recursion induction" for proving equivalence, and a comparison of his system with others in recursive function theory and programming.

In his first presentation of his system in 1960, McCarthy had used a variation of LISP as a metalanguage. He then introduced the lambda calculus, to which he had earlier turned when seeking a notation that allowed the distribution of a function over a list with an indeterminate number of arguments. Concerned primarily with LISP as a working language and satisfied with its metatheoretical qualities, McCarthy did not pursue the further reduction of LISP to the lambda calculus; indeed, he adopted Nathaniel Rochester's concept of *label* as a means of circumventing both the complicated expression and the self-applicative function required by recursive definition in the pure notation. Others, most notably Peter J. Landin, did look to the lambda calculus itself as a metalanguage for programs, seeing several advantages in it. It made the scope of bound variables explicit and thus prevented clashes of scope during substitution (that is one reason why Church designed it). Its rules and procedures for reduction to normal form made it possible to show that two different expressions were equivalent in the sense of having the same result when applied to the same arguments. Moreover, it was type-free, treating variables and functions as equally abstractable entities.[9]

The first property clarified the complexities of evaluating the arguments of a function when their variables have the same name as those of the function. The second property provided analytical insight into the structure of functions, showing how they were constructed from basic functions and allowing transformations among them. In McCarthy's system, it underlay the technique of recursive induction. For example, let integer addition be defined recursively by the conditional equation $m + n = (n = 0 \to m, T \to m' + n^-)$, where m' is the successor of m and n^- is the predecessor of n. To show that $(m + n)' = m' + n$, let $g(m,$

$n) = (m + n)' = (n = 0 \rightarrow m, \mathrm{T} \rightarrow m' + n^-)' = (n = 0 \rightarrow m', \mathrm{T} \rightarrow (m' + n^-)') = (n = 0 \rightarrow m', \mathrm{T} \rightarrow g(m',n^-))$, and $h(m,n) = m' + n = (n = 0 \rightarrow m', \mathrm{T} \rightarrow (m')' + n^-) = (n = 0 \rightarrow m', \mathrm{T} \rightarrow h(m',n^-))$. Whence g and h both satisfy the relation $f = \lambda m.\lambda n.(n = 0 \rightarrow m', \mathrm{T} \rightarrow f(m',n^-))$ when substituted for f and hence are equal.[10]

The third property opened a link to the particular nature of the stored-program computer and thus fitted McCarthy's rephrasing of his expectations in a second paper, "Towards a Mathematical Science of Computation," delivered at IFIP 62. The entities of computer science consist of "problems, procedures, data spaces, programs representing procedures in particular programming languages, and computers." Once distinguished from problems, defined by the criteria of their solution, the construction of complex procedures from elementary ones could be understood in terms of the well-established theory of computable functions. However, there was no comparable theory of the representable data spaces on which those procedures operate. Similarly, while the syntax of programming languages had been formalized, their semantics remained to be studied. Finally, despite the fact that computers are finite automata, "Computer science must study the various ways elements of data spaces are represented in the memory of the computer and how procedures are represented by computer programs. From this point of view, most of the current work on automata theory is beside the point."

McCarthy did not persuade many of his leading American colleagues, who doubted the need for, and feasibility of, a formal semantics, but on this last point he found an ally in Strachey, for whom Landin had been working and who built his contribution to the 1964 Working Conference on Formal Description Languages in Vienna precisely on the question of what goes on in the memory (store) of a computer and on the "essentially computer-oriented" operations of assignment and transfers of control that go on there. In "Toward a Formal Semantics," Strachey worked from the model of a computer's memory as a finite set of N objects, well ordered in some way by a mapping that assigns to each of them a *name*, or *L-value*. Each object is itself a binary array, which may be viewed as the *value*, or *R-value* associated with the name. A program consists of a sequence of operations applied to names and values to produce values associated with names; in other words, a mapping of names and values into names and values. However the operations are defined abstractly; they reduce to the instruction set of the processor. In principle, one should be able to treat a program as a mathematical object and analyze its structure.

That structure cannot be entirely abstract or syntactical, at least not if it is to meet the most basic requirements of real programming. As an analysis of the assignment command shows, it is necessary to distinguish between the *L-value* and *R-value* of an expression. That is, the command $\varepsilon_1 := \varepsilon_2$ requires that the expression on the left be interpreted as a name and that on the right as a value; the two expressions require different evaluations. While one could make that evaluation trivial by restricting the command to allow only primitive names on the left, doing so would sacrifice such features as list-processing in LISP and Strachey's own CPL. Moreover, the value of a name may extend beyond a binary array to include an area of memory, as in the case, peculiar to the stored-program computer, where it contains executable code. Thus, expressions such as "(if $x > 0$ then *sin* else *cos*)(x)," meaningless to the mathematical eye, make sense computationally: the variable x and the procedures for *sin* and *cos* are equally valid values of their names.

To capture the structure of this model of memory, Strachey introduced two operators, "loading" and "updating," which retrieve and store the *R-values* associated with *L-values*. Symbolically, let α denote an *L-value* and β its associated *R-value*, and let σ denote the "content of the store" or the total set of *R-values* at any given moment. Then "loading," denoted by C, will be a function of α, which when applied to σ yields β, that is, $\beta = (C(\alpha))(\sigma)$. "Updating," denoted U, produces a new content σ' through the operation $(U(\alpha))(\beta',\sigma)$, where β' is a value compatible with β. Hence, if one treats the "natural" result of an expression ε as its *L-value*, expressed symbolically as $\alpha = L(\varepsilon,\sigma)$, then its *R-value* can be obtained by means of the loading operator: $\beta = R(\varepsilon,\sigma) = (C(L(\varepsilon,\sigma)))(\sigma)$. Introduced as functions into the descriptive expressions of a language, Strachey argued, these operators provided a specification of how the results of the expressions should be treated at the level of the computer.

Drawing on Landin's work, Strachey embedded the L and R functions into the λ-calculus, which he and his collaborators used as a metalanguage for specifying and analyzing the semantics of programming languages. Although they called the enterprise mathematical, it had no underlying mathematical structure to serve as model for the formal system. As Scott insisted when he and Strachey met in Vienna in 1968, their analysis amounted to no more than a translation of the object language into the metalanguage. How the data types and functions of the language were to be constructed mathematically remained an open question. Scott's criticism of Strachey echoed Anil Nerode's reaction to McCarthy's approach. There was no mathematics in it.[11]

Scott had been working in various areas of logic during the '60s, having concluded that none of the areas of theoretical computer science was heading in promising directions. He had gradually formed his own idea of where and how the mathematics entered the picture. He sought the middle ground in the tension inherent in applied mathematics. The mathematics moves in the direction of ever-greater abstraction from the intended application. Yet, the application sets the conditions for the abstraction. The mathematical model must maintain contact with the physical model. The test of practicality always looms over the effort. A mathematical theory of computation addressed to understanding programs has to connect the abstract model to the concrete machine, Scott argued in 1970: "an *adequate* theory of computation must not only provide the abstractions (what is computable) but also their 'physical' realizations (how to compute them)." The means of realization had been known for some time, he added; what was needed were the abstractions, which could expose the structure of a programming language. "Now it is often suggested that the meaning of the language resides in one particular compiler for it. But that idea is wrong: the 'same' language can have many 'different' compilers. The person who wrote one of these compilers obviously had a (hopefully) clear understanding of the language to guide him, and it is the purpose of mathematical semantics to make this understanding 'visible.' This visibility is to be achieved by abstracting the central ideas into mathematical entities, which can then be 'manipulated' in the familiar mathematical manner."[12]

The mathematical entities derived from the physical structure of the computer. Mathematical semantics concerned *data types* and the *functions* that map them from one to another. The spaces of those functions also form data types. The finite structure of the computer means that some finite *approximation* is needed for functions, which are by nature infinite objects (e.g., mappings of integers to integers). Because computers store programs and data in the same memory, programming languages allowed unrestricted procedures which could have unrestricted procedures as values; in particular a procedure could be applied to itself. "To date," Scott claimed,

> no mathematical theory of functions has ever been able to supply conveniently such a free-wheeling notion of function except at the cost of being inconsistent. The main *mathematical* novelty of the present study is the creation of a proper mathematical theory of functions which accomplishes these aims (consistently!) and which can be used as the basis for the *metamathematical* project of providing the "correct" approach to semantics.

One did not need unrestricted procedures to appreciate the problems posed by the self-application facilitated by the design of the computer. Following Strachey, consider the structure of computer memory, representing it mathematically as a mapping of contents to locations. That is, state σ is a function mapping each element ℓ of the set L of locations to its value $\sigma(\ell)$ in V, the set of allowable values. A command effects a change of state; it is a function γ from the set of states S into S. Storing a command means that γ can take the form $\sigma(\ell)$, and hence $\sigma(\ell)(\sigma)$ should be well defined. Yet, "this is just an insignificant step away from the self-application problem $p(p)$ for 'unrestricted' procedures p, and it is just as hard to justify mathematically."[13]

Recent work on interval arithmetic suggested that one might seek justification through a partial ordering of data types and their functions based on the notion of "approximation" or "informational content." With the addition of an undefined element as "worst approximation" or "containing no information," the data types formed a complete lattice, and monotonic functions of them preserved the lattice. They also preserved the limits of sequences of partially ordered data types and hence were continuous. Scott showed that the least upper bound of the lattice, considered as the limit of sequences, was therefore the least fixed point of the function and was determined by the fixed-point operator of the λ-calculus. Hence self-applicative functions of the sort needed for computers had a consistent mathematical model. And so too, by the way, did the λ-calculus for the first time in its history.

Scott's lattice-theoretical model established a rigorous mathematical foundation for the program Strachey had proposed in 1964. Together they wrote "Toward a Mathematical Semantics for Computer Languages," which "covers much the same ground as Strachey ["Toward a Formal Semantics"], but this time the mathematical foundations are secure. It is also intended to act as a bridge between the formal mathematical foundations and their application to programming languages by explaining in some detail the notation and techniques we have found to be most useful." Mathematical semantics formed another sort of bridge as well. It led back to the body of algebraic structures that had provided previous models of computing, but it now spanned the gap between finite-state machines and Turing machines (in the equivalent form of the lambda calculus) by taking account of the random-access, stored program device that embodied them both.[14]

By 1970 computer science had assumed a shape recognized by both the mathematical and the computing communities, and it could point to both applications and mathematical elegance. Yet, it took the form more of a family of loosely related research agendas than of a coherent general theory validated by empirical results. So far, no one mathematical model had proved adequate to the diversity of computing, and the different models were not related in any effective way. What mathematics one used depended on what questions one was asking, and for some questions no mathematics could account in theory for what computing was accomplishing in practice.

CHAPTER TWELVE

The Structures of Computation and the Mathematical Structure of Nature

N 1984 SPRINGER-VERLAG added to its series Undergraduate Texts in Mathematics a work with the seemingly oxymoronic title *Applied Abstract Algebra*, by Rudolf Lidl and Gunter Pilz. To the community of practitioners the work may have seemed, as the cliché puts it, "long overdue"; indeed, it was one of a spate of similar works that appeared at the time, evidently filling a perceived need. To the historian of modern mathematics, it appears remarkable on several counts, not least that it was meant for undergraduates. Less than fifty years earlier, Garrett Birkhoff's father had asked him what use his new theory of lattices might possibly have, and many non-algebraists had wondered the same thing about the subject as a whole. *Applied Abstract Algebra* opens with a chapter on lattices, which form a recurrent theme throughout the book. Subsequent chapters present rings, fields, and semigroups. Interspersed among those topics and motivating their organization are the applications that give the book its title and purpose. They include switching circuits, codes, cryptography, automata, formal languages: in short, problems arising out of computing or intimately related to it.

Although the particular topics and treatments differed, Lidl and Pilz were following the lead of Birkhoff himself, who in 1970 together with Thomas Bartee had assembled *Modern Applied Algebra*, and explained its purpose right at the outset:

> The name "modern algebra" refers to the study of algebraic systems (groups, rings, Boolean algebras, etc.) whose elements are typically *non-numerical*. By contrast, "classical" algebra is basically concerned with algebraic equations

or systems of equations whose symbols stand for real or complex numbers. Over the past 40 years, "modern" algebra has been steadily replacing "classical" algebra in American college curricula.[1]

The past 20 years have seen an enormous expansion in several new areas of technology. These new areas include digital computing, data communication, and radar and sonar systems. Work in each of these areas relies heavily on modern algebra. This fact has made the study of modern algebra important to applied mathematicians, engineers, and scientists who use digital computers or who work in the other areas of technology mentioned above.[2]

That is, by 1970 the most abstract notions of twentieth-century mathematics had evidently found application in the most powerful technology of the era, indeed the defining technology of the modern world. Indeed, the development of computing since 1970, both theoretical and applied, has repeatedly reinforced the relationship, as one can see from titles like *The Algebra of Programming*; *Categories, Types, and Structures: An Introduction to Category Theory for the Working Computer Scientist*; and *Basic Category Theory for Computer Scientists*, to name just a few.[3]

The Problem of Applied Science and Applied Mathematics

That strikes me as a remarkable development, all the more so for a discussion that followed a paper I gave a couple of years ago at a conference on the history of software. Computer scientists and software engineers in the audience objected to my talking about the development of theoretical computer science as a mathematical discipline under the title "Software as Science—Science as Software." The matters at issue were not technical but professional, philosophical, and historical. My critics disagreed with me—and with each other—over whether software (i.e., computer programs and programming) could be the subject of a science, whether that science was mathematics, and whether mathematics could be considered a science at all.[4]

There was much common wisdom in the room, expressed with the certainty and conviction that common wisdom affords. Two arguments in particular are pertinent here. First, the computer is an artifact, not a natural phenomenon, and science is about natural phenomena. Second, as a creation of the human mind, independent of the physical world, mathematics is not a science. It is at most a tool for doing science.

Modern technoscience undercuts the first point. How does one distinguish between nature and artifact when we rely on artifacts to produce or afford access to the natural phenomena, as with accelerators, electron microscopes, and space probes—all, one might add, mediated by software?

We need not wait until the twentieth century to find technology intertwined with science. In insisting that "nature to be commanded must be obeyed," Francis Bacon placed nature and art on the same physical and epistemological level. Artifacts work by the laws of nature, and by working reveal those laws. Only with the development of thermodynamics, which began with the analysis of steam engines already at work, did we "discover" that world is a heat engine subject to the laws of entropy. A century later came information theory, the analysis of communications systems arising from the problems of long-distance telephony, again already in place. Information could be related via the measure of entropy to thermodynamics, and the phenomenal world began to contain information as a measure of its order and structure. According to Stephen Hawking, quantum mechanics, relativity, thermodynamics, and information theory all meet on the event horizons of black holes. There's a lot of physics here, but also a lot of mathematics, and a lot of artifacts lying behind both.[5]

Now, with the computer, nature has increasingly become a computation. DNA is code, the program for the process of development; the growth of plants follows the recursive patterns of formal languages, and biochemical molecules implement automata. Stephen Wolfram's *A New Kind of Science* is only the most extreme assertion of an increasingly widespread view, expressed with equal fervor by proponents of artificial life. Although the computational world may have begun as a metaphor, it is now acquiring the status of metaphysics, thus repeating the early modern transition from the metaphor of "machine of the world" to the metaphysics of matter in motion. The artifact as conceptual scheme is deeply, indeed inseparably, embedded in nature, and the relationship works both ways, as computer scientists turn to biological models to address problems of stability, adaptability, and complexity.[6]

Embedded too is the mathematics that has played a central role in the articulation of many of these models of nature—thermodynamical, informational, and computational—not simply by quantifying them but also, and more importantly, by capturing their structure and even filling it out. The calculus of central-force dynamics revealed by differentiation the Coriolis force before Coriolis identified it, and Gell-Mann's and Ne'eman's Ω^- baryon served as the tenth element needed to complete a group representation before it appeared with its predicted properties in the Brookhaven accelerator. One can point to other examples in between. Applied to the world as models, mathematical structures have captured its workings in ways thinkers since the seventeenth century have found uncanny or, as Eugene Wigner put it, "unreasonably effective."[7]

Is mathematics a science of the natural world, or can it be? Computational science puts the question in a new light. Science is not about nature, but about how we represent nature to ourselves. We know about nature through the models we build of it, constructing them by abstraction from our experience, manipulating them physically or conceptually, and testing their implications back against nature. How we understand nature reflects, first, the mapping between the observable or measurable parameters of a physical system and the operative elements of the model and, second, our capacity to analyze the structure of the model and the transformations of which it is capable. In the physical sciences, the elements have been particles in motion or distributions of force in fields described in differential equations; in the life sciences, they have been organisms arranged in a taxonomy or gathered in statistically distributed populations. The scope and power of these models have depended heavily on the development of mathematical techniques to analyze them and to derive new structures and relations within them. Developed initially as a tool for solving systems of equations that were otherwise intractable, computers have evolved to provide a means for a new kind of modeling and thus a new kind of understanding of the world, namely, dynamic simulation, which is what we generally mean when we talk of "computational science." But with the new power come new problems, first, to define the mapping that relates the operative elements of the simulation to what we take to be those of the natural system and, second, to develop the mathematical tools to analyze the dynamic behavior of computational processes and to relate their structures to one another.

In short, understanding nature appears to be coming down to understanding computers and computation. Whether computer science, broadly conceived, should be or even could be a wholly mathematical science remains a matter of debate among practitioners, on which historians must be agnostic. What matters to this historian is that from the earliest days, leading members of the community of computer scientists, as measured by citations, awards, and honors, have turned to mathematics as the foundation of computer science. I am not stating a position but reporting a position taken by figures who would seem to qualify as authoritative. A look at what current computing curricula take to be fundamental theory confirms that position, and developments in computational science reinforce it further.

So, let me repeat my critics' line of thinking: computation is what computers do, the computer is an artifact, and computation is a mathematical concept. So how could computation tell us anything about nature? Or rather, to take the historian's view, how has it come about that computation is both a mathematical subject and an increasingly dominant model of nature? And how has that parallel development redefined the relations among technology, mathematics, and science?

Applied Mathematics and Its History

The answer is tied in part to a shift in the meaning of "applied mathematics" that took place during the mid-twentieth century. It shifted from putting numbers on theory to modeling as abstraction of structure. The shift is related to Birkhoff's differentiation between "classical" and "modern" algebra: the latter redirects attention from the objects being combined to the combinatory operations and their structural relations. Thus homomorphisms express the equivalence of structures arising from different operations, and isomorphism expresses the equivalence of structures differing only in the particular objects being combined. From functional analysis to category theory, algebra moved upward in a hierarchy of abstraction in which the relations among structures of one level became the objects of relations at the next. As Herbert Mehrtens has pointed out, mathematics itself was following in the direction Hilbert had indicated at the turn of the twentieth century, beholden no longer to physical ontology or to philosophical grounding.[8]

The new kind of modeling followed that pattern and fitted well with concurrent shifts in the relation between mathematics and physics occasioned by the challenges of relativity and quantum mechanics, which pushed mathematical physics onto new ground, moving it from the mathematical expression of physical structures to the creation of mathematical structures open to, but not bound by, physical interpretation. A shift of technique accompanied the shift of ground. As Bruna Ingrao and Giorgio Israel observe in their history of equilibrium theory in economics,

> . . . one went from the theory of individual functions to the study of the collectivity of functions (or functional spaces); from the classical analysis based on differential equations to abstract functional analysis, whose techniques referred above all to *algebra* and to *topology*, a new branch of modern geometry. The *mathematics of time*, which had originated in the Newtonian revolution and developed in symbiosis with classical infinitesimal calculus, was defeated by a *static and atemporal* mathematics. A rough and approximate description of the ideal of the new mathematics would be fewer differential equations and more inequalities. In fact, the new mathematics, of which von Neumann was one of the leading authors, was focused entirely upon techniques of functional analysis, measurement theory, convex analysis, topology, and the use of fixed-point theorems. One of von Neumann's greatest achievements was unquestionably that of having grasped the cen-

tral role within modern mathematics of fixed-point theorems, which were to play such an important part in economic equilibrium theory.[9]

Ingrao and Israel echoed what Marcel P. Schützenberger (about whom more below) had earlier signaled as the essential achievement of John von Neumann's theory of games, which he characterized as "the first conscious and coherent attempt at an axiomatic study of a psychological phenomenon."

> In von Neumann's theory there is no reference to a physical substrate of which the equations will serve to symbolize the mechanism, nor are there introduced any but the minimum of psychological or sociological observations, which more or less fragmentary analytic schemes will seek to approximate. On the contrary, from the outset, the most general postulates possible, chosen certainly on the basis of extra-mathematical considerations that legitimate them a priori, but accessible in a direct way to mathematical treatment. These are mathematical hypotheses with psychological content—not mathematizable psychological data.[10]

Such an approach, Schützenberger added, required that von Neumann fashion new analytical tools from abstract algebra, including group theory and the theory of binary relations, thus making his application of mathematics to psychology at the same time "a new chapter in mathematics." The physical world thus became an interpretation, or semantic model, of an essentially uninterpreted, abstract mathematical structure.

Von Neumann's Agendas: The Formation of Theoretical (Mathematical) Computer Science

John von Neumann first turned to the computer as a tool for applied mathematics in the traditional sense. Numerical approximation had to suffice for what analysis could not provide when faced with non-linear partial differential equations encountered in hydrodynamics. But he recognized that the computer could be more than a high-speed calculator. It allowed a new form of scientific modeling, and it constituted an object of scientific inquiry in itself. That is, it was not only a means of doing applied mathematics of the traditional sort but a medium for applied mathematics of a new sort. Hence, his work on the computer took him in two directions, namely, numerical analysis for high-speed computation (what became scientific computation) and the theory of automata as "artificial organisms" (fig. 6).[11]

In proposing new lines of research into automata and computation, von Neumann knew that it would require new mathematical tools, but he was not sure of where to look for those tools.

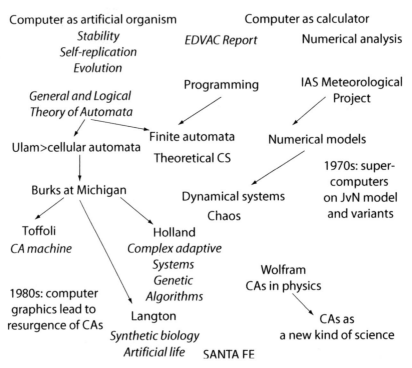

Figure 6   The agenda leading from John von Neumann to the Santa Fe Institute.

This is a discipline with many good sides, but also with certain serious weaknesses. This is not the occasion to enlarge upon the good sides, which I certainly have no intention to belittle. About the inadequacies, however, this may be said: Everybody who has worked in formal logic will confirm that it is one of the technically most refractory parts of mathematics. The reason for this is that it deals with rigid, all-or-none concepts, and has very little contact with the continuous concept of the real or of the complex number, that is, with mathematical analysis. Yet analysis is the technically most successful and best-elaborated part of mathematics. Thus formal logic is, by the nature of its approach, cut off from the best cultivated portions of mathematics, and forced onto the most difficult part of the mathematical terrain, into combinatorics.

The theory of automata, of the digital, all-or-none type, as discussed up to now, is certainly a chapter in formal logic. It will have to be, from the mathematical point of view, combinatory rather than analytical.[12]

At the time von Neumann wrote this, there were two fields of mathematics that dealt with the computer: mathematical logic and the Boolean algebra of circuits. The former treated computation at its most abstract and culminated with what could not be computed. The latter dealt with how to compute at the most basic level. Von Neumann was looking for something in the middle, and that is what emerged over the next fifteen years. It came, as von Neumann suspected, from combinatory mathematics rather than from analysis, although not entirely along the lines he expected. His own research focused on "growing" automata (in the form of cellular automata) and the question of self-replication, a line of inquiry picked up by Arthur Burks and his Logic of Computers Group at the University of Michigan. This is an important subject still awaiting a historian.

The story of how the new subject assumed an effective mathematical form is both interesting and complex, and I have laid it out in varying detail in several articles. Let me here use a couple of charts to describe the main outline and to highlight certain relationships, among them what I call a convergence of agendas.[13]

Agendas

At the heart of a discipline lies an agenda, a shared sense among its practitioners of what is to be done: the questions to be answered, the problems to be solved, the priorities among them, the difficulties they pose, and where the answers will lead. When one practitioner asks another, "What are you working on?," it is their shared agenda that gives professional meaning to both the question and the response. Learning a discipline means learning its agenda and how to address it, and one acquires standing by solving problems of recognized importance and ultimately by adding fruitful questions and problems to the agenda. Indeed, the most significant solutions are precisely those that expand the agenda by posing new questions. Disciplines may cohere more or less tightly around a common agenda, with subdisciplines taking responsibility for portions of it. There may be some disagreement about problems and priorities. But there are limits to disagreement, beyond which it can lead to splitting, as in the separation of molecular biology from biochemistry and from biology.

Most important for present purposes, the agendas of different disciplines may intersect on what come to be recognized as common problems, viewed at first from different perspectives and addressed by different methods but gradually forming an autonomous agenda with its own body of knowledge and practices specific to it. Theoretical computer science presents an example of how such an intersection of agendas generated a new agenda and with it a new mathematical discipline.[14]

Automata and Formal Languages

As the first chart suggests, the theory of automata and formal languages resulted from the convergence of a range of agendas on the common core of the correspondence between the four classes of Noam Chomsky's phrase-structure grammars and four classes of finite automata. To the first two classes correspond as well algebraic structures which capture their properties and behavior. The convergence occurred in several stages. First, the deterministic finite automaton took form as the common meeting ground of electrical engineers working on sequential circuits and of mathematical logicians concerned with the logical capabilities of nerve nets as originally proposed by Warren McCulloch and Walter Pitts and developed further by von Neumann. Indeed, Stephen Kleene's investigation of the McCulloch-Pitts paper was informed by his work at the same time on the page proofs for von Neumann's *General and Logical Theory of Automata*, which called his attention to the finite automaton. E. F. Moore's paper on sequential machines and Kleene's on regular events in nerve nets appeared together in *Automata Studies* in 1956, although both had been circulating independently for a time before that. Despite their common reference to finite automata, the two papers took no account of each other. They were brought together by Michael Rabin and Dana Scott in their seminal paper on "Finite Automata and Their Decision Problems," which drew a correspondence between the states of a finite automaton and the equivalence classes of the sequences recognized by them (fig. 7).[15]

These initial investigations were couched in a combination of logical notation and directed graphs, which suited their essentially logical concerns. Kleene did suggest at the end of his article that "the study of a set of objects $a_1, \ldots, a_r$ under a binary relation R, which is at the heart of the above proof [of his main theorem], might profitably draw on some algebraic theory." Rabin and Scott in turn mentioned that the concatenated tapes of an automaton form a free semigroup with unit. But neither of the articles pursued the algebra any farther. The algebraic theory came rather from another quarter. In 1956 at the seminal Symposium on Information Theory held at MIT, Marcel P. Schützenberger contributed a paper, "On an Application of Semi-Group Methods to Some Problems in Coding." There he showed that the structure of the semigroup captured the essentially sequential nature of codes in transmission, and the semigroup equivalent of the word problem for groups expressed the central problem of recognizing distinct subsequences of symbols as they arrive. The last problem in turn is a version of the recognition problem for finite

Automata and Formal Languages

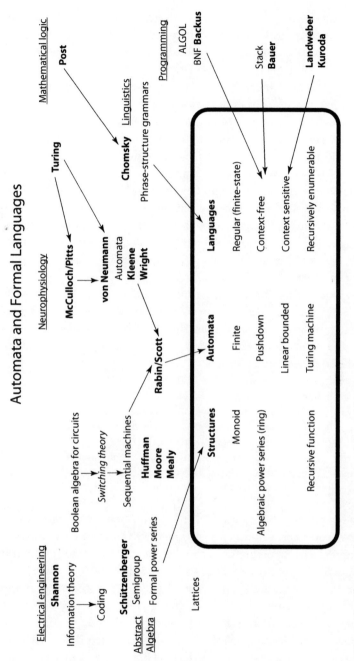

Figure 7 The agendas of automata and formal languages.

automata as defined and resolved by Kleene and then Rabin and Scott. Turning from codes to automata in a series of papers written between 1958 and 1962, Schützenberger established the monoid, or semigroup with unit, as the fundamental mathematical structure of automata theory, demonstrating that the states of a finite-state machine, viewed as a homomorphic image of the equivalence classes of strings indistinguishable by the machine, form a monoid and that the subset of final states is a closed homomorphic image of the strings recognized by the machine. Reflecting the agenda of Bourbaki's algebra, he then moved from monoids to semi-rings by expressing the sequences recognized by an automaton as formal power series in non-commutative variables. On analogy with power series in real variables, he distinguished among "rational" and "algebraic" power series and explored various families of the latter. Initially he identified "rational" power series with finite automata, and hence finite-state languages, but could point to no language or machine corresponding to the "algebraic" series.[16]

That would come through his collaboration with linguist Noam Chomsky, who at the same 1956 symposium introduced his new program for mathematical linguistics with a description of "three models of language" (finite-state, phrase-structure, and transformational). Chomsky made his scientific goal clear at the outset:

> The grammar of a language can be viewed as a theory of the structure of this language. Any scientific theory is based on a certain finite set of observations and, by establishing general laws stated in terms of certain hypothetical constructs, it attempts to account for these observations, to show how they are interrelated, and to predict an indefinite number of new phenomena. A mathematical theory has the additional property that predictions follow rigorously from the body of theory.[17]

Following work with George Miller on finite-state languages viewed as Markov sequences, Chomsky turned in 1959 to those generated by more elaborate phrase-structure grammars, the context-free, context-sensitive, and recursively enumerable languages. The technique here stemmed from the production systems of Emil Post by way of Paul Rosenbloom's treatment of concatenation algebras. Of these, the context-free languages attracted particular interest for both theoretical and practical reasons. Theoretically, Schützenberger determined in 1961 that they corresponded to his algebraic power series. Practically, Ginsburg and Rice showed that the syntax of portions of the newly created programming language ALGOL constituted a context-free grammar. They did so by transforming the Backus Normal Form description of those portions into set-theoretical equations among monotonically increasing functions forming a lattice and hence, by Alfred Tarski's well-known result, having a minimal fixed point, which constituted the language generated by the grammar. At the same time, Anthony Oettinger brought together a variety of studies on machine translation and syntactic analysis converging on the concept of the stack, or last-in-first-out list, also featured in the work of Klaus Samelson and Friedrich Bauer on sequential formula translation. Added to a finite automaton as a storage tape, it formed the pushdown automaton. In 1962, drawing on Schützenberger's work on finite transducers, Chomsky showed that the context-free languages were precisely those recognized by a non-deterministic pushdown automaton. Their work culminated in a now-classic essay, "The Algebraic Theory of Context-Free Languages," and in Chomsky's chapter on formal grammars in the *Handbook of Mathematical Psychology*.[18]

By the mid-1960s, Schützenberger's work had established the monoid and formal power series as the fundamental structures of the mathematical theory of automata and formal languages, laying the foundation for a rapidly expanding body of literature on the subject and, as he expected of the axiomatic method, for a new field of mathematics itself. As his student, Dominique Perrin, has noted, the subsequent development of the subject followed two divergent paths, marked on the one hand by works such as Samuel Eilenberg's *Automata, Languages, and Machines* ("la version la plus mathématisée des automates") and on the other by the textbooks of A. V. Aho, John Hopcroft, and Jeffrey Ullman, in which the theory is adapted to the more practical concerns of practitioners of computing. In that they follow the "personal bias" of Rabin, who opined in 1964 that

> . . . whatever generalization of the finite automaton concept we may want to consider, it ought to be based on some intuitive notion of machines. This is probably the best guide to fruitful generalizations and problems, and our intuition about the behavior of machines will then be helpful in conjecturing and proving theorems.[19]

The subject would be fully mathematical, but it would retain its roots in the artifact that had given rise to it.

Formal Semantics

From a mathematical point of view, formal semantics is the story of how the λ-calculus acquired a mathematical structure and of how universal algebra and category theory became useful in analyzing its untyped and typed forms, respectively. The story begins, by general acknowledgment, with the work of John McCarthy on the theoretical implications of LISP, as set forth in his 1960 article "Recursive Functions of Symbolic Expressions and Their Computation by Machine" and then in two papers, "A Basis for a Mathematical Theory of Computation" and "Towards a Mathematical Science of Computation." In the article, he outlined how the development of LISP had led to a machine-independent formalism for defining functions recursively and to a universal function that made the formalism equivalent to Turing machines. To express functions as object in themselves, distinct both from their values and from forms that constitute them, McCarthy turned to Alonzo Church's λ-calculus, preferring its explicitly bound variables to Haskell Curry's combinatory logic, which avoided them only at the price of "lengthy, unreadable expressions for any interesting combinations." However, Church's system had its drawbacks, too. "The λ-notation is inadequate for naming functions defined recursively," McCarthy noted. To remedy the problem, he introduced (on the suggestion of Nathaniel Rochester) a notation *label* to signify that the function named in the body of the λ-expression was the function denoted by the expression. It was a stop-gap measure, and McCarthy returned to the problem in his subsequent articles. It proved to be the central problem of the λ-calculus approach to semantics (fig. 8).[20]

The two articles that followed built on the system of symbolic expressions, in which McCarthy now saw the foundations for a mathematical theory of computation. In "Basis" he set out some of the goals of such a theory and thus an agenda for the field. It should aim:

> [1] to develop a universal programming language . . . [2] to define a theory of the equivalence of computational processes . . . [3] to represent algorithms by symbolic expressions in such a way that significant changes in the behavior represented by the algorithms are represented by simple changes in the symbolic expressions . . . [4] to represent computers as well as computations in a formalism that permits a treatment of the relation between a computation and the computer that carries out the computation . . . [5] to give a quantitative theory of computation . . . analogous to Shannon's measure of information.[21]

The article focused on two of these goals, setting out the formalism of symbolic expressions and introducing the method of recursion induction as a means of determining the equivalence of computational processes.

In the second paper, presented at IFIP 62, McCarthy defined his goals even more broadly, placing his mathematical science of computation in the historical mainstream:

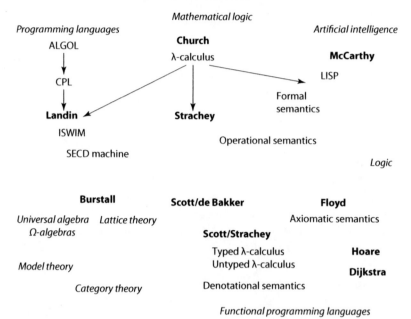

Formal Semantics

Figure 8   The agendas of semantics.

In a mathematical science, it is possible to deduce from the basic assumptions, the important properties of the entities treated by the science. Thus, from Newton's law of gravitation and his laws of motion, one can deduce that the planetary orbits obey Kepler's laws.

The entities of computer science, he continued, are "problems, procedures, data spaces, programs representing procedures in particular programming languages, and computers." Especially problematic among them were data spaces, for which there was no theory akin to that of computable functions; programming languages, which lacked a formal semantics akin to formal syntax; and computers viewed as finite automata. "Computer science," he admonished, "must study the various ways elements of data spaces are represented in the memory of the computer and how procedures are represented by computer programs. From this point of view, most of the current work on automata theory is beside the point."[22]

Whereas the earlier articles had focused on theory, McCarthy turned here to the practical benefits of a suitable mathematical science. It would be a new sort of mathematical theory, reaching beyond the integers to encompass all domains definable computationally and shifting from questions of unsolvability to those of constructible solutions. It should ultimately make it possible to prove that programs meet their specifications, as opposed simply (or not so simply) to debugging them. . . . There was no question in McCarthy's mind that computing was a mathematical subject. The problem was that he did not yet have the mathematics. In particular, insofar as his formalism for recursive definition rested on the λ-calculus, it lacked a mathematical model.[23]

McCarthy's new agenda for the theory of computation did not enlist much support at home, but it struck a resonant chord in Europe, where work was under way on the semantics of ALGOL 60 and PL/1. Christopher Strachey had learned about the λ-calculus from Roger Penrose in 1958 and had engaged Peter J. Landin to look into its application to formal semantics. In lectures at the University of London Computing Unit in 1963, Landin set out "A λ-Calculus Approach" to programming, aimed at "introduc[ing] a technique of semantic analysis and . . . provid[ing] a mathematical basis for it." Here Landin followed McCarthy's lead but diverged from it in several respects, in particular by replacing McCarthy's LISP-based symbolic expressions with "applicative expressions" (AEs) derived directly from the λ-calculus but couched in "syntactic sugar" borrowed from the new Combined Programming Language (CPL) that Strachey and a team were designing for new machines at London and Cambridge. More important, where McCarthy had spoken simply of a "state vector," Landin articulated it into a quadruple (S,E,C,D) of lists and list-structures consisting of a stack, an environment, a control, and a dump, and specified in terms of an applicative expression the action by which an initial configuration is transformed into a final one.[24]

But Landin was still talking about a notional machine, which did not satisfy Strachey, who embraced the λ-calculus for reasons akin to those of Church in creating it. Where Church sought to eliminate free variables and the need they created for an auxiliary, informal semantics, Strachey wanted to specify the semantics of a programming language formally without the need for an informal evaluating mechanism. In "Towards a Formal Semantics," he sought to do so, and to address the problem of assignment and jumps, by representing the store of a computer as a set of "R-values" sequentially indexed by a set of "L-values." The concept (suggested by Rod Burstall) reflected the two different ways the components of an assignment are evaluated by a compiler: in $X := Y$, the X is resolved into a location in memory into which the contents of the location designated by Y is stored. Strachey associated with each L-value α two operators, $C(\alpha)$ and $U(\alpha)$, respectively denoting the loading and updating of the "content of the store" denoted by σ. C(α) applied to σ yields the R-value β held in location α, and U(α) applied to the pair (β′,σ) produces a new σ′ with β′ now located at α. One can now think of a command as an operator θ that changes the content of the store by updating it. Grafting these operations onto Landin's λ-calculus approach to evaluating expressions, Strachey went on to argue that one could then reason about a program in terms of these expressions alone, without reference to an external mechanism.[25]

Strachey's argument was a sketch, rather than a proof, and its structure involved a deep problem connected with the lack of a mathematical model for the λ-calculus, as Dana Scott insisted when he and Strachey met in Vienna in August 1969 and undertook collaborative research at Oxford in the fall. As Dana Scott phrased it, taking S as a set of states σ, L as the set of locations (i.e., L-values) l, and V as the set of (R-)values, one can rewrite the loading operator in the form σ(l) V. A command γ is then a mapping of S into S. But γ is also located in the store and hence in V; that is, γ = σ(l) for some l. Hence, the general form of a command may be written as γ(σ) = (σ(l))(σ), which lies "an insignificant step away from . . . $p(p)$." Scott continued:

> To date, no mathematical theory of functions has ever been able to supply conveniently such a free-wheeling notion of function except at the cost of being inconsistent. The main *mathematical* novelty of the present study is the creation of a proper mathematical theory of functions which accomplishes these aims (consistently!) and which can be used as the basis for the *metamathematical* project of providing the "correct" approach to semantics.

The mathematical novelty, which took Scott by surprise, was finding that abstract data types and functions on them could be modeled as continuous lattices, which thereby constituted a model for the type-free λ-calculus; by application of Tarski's theorem, the self-reference of recursive definitions, as represented by the Y-operator, resolved into the least fixed point of the lattice.[26]

By the mid-1960s proponents of formal semantics had articulated two goals. First, they sought a metalanguage in which the meaning of programming languages could be specified with the same precision with which their syntax could be defined. That is, a compiler should be provably able to translate the operational statements of a language into machine code as unambiguously as a parser could recognize their grammar. Second, the metalanguage should allow the mathematical analysis of programs to establish such properties as equivalence, complexity, reducibility, and so on. As Rod Burstall put it in 1968,

The aims of these semantic investigations are twofold. First they give a means of specifying programming languages more precisely, extending the syntactic precision which was first achieved in the Algol report to murkier regions which were only informally described in English by the Algol authors. Secondly they give us new tools for making statements *about* programs, particularly for proving that two dissimilar programs are in fact equivalent, a problem which tends to be obscured by the syntactic niceties of actual programming languages.[27]

Ultimately, the two goals came down to one. In the end, argued Dana Scott in the paper just mentioned,

> the program still must be run on a machine—a machine which does not possess the benefit of abstract human understanding, a machine that must operate with finite configurations. Therefore, a mathematical semantics, which will represent the first major segment of the complete, rigourous definition of a programming language, must lead naturally to an operational simulation of the abstract entities, which—if done properly—will establish the practicality of the language, and which is necessary for a full presentation.[28]

For the time being, mathematical discourse and working programs were different sorts of things, whatever the imperatives of theory or the long-term aspirations of artificial intelligence.

The difference is evident in McCarthy's two major papers. "Basis" is about a mathematical theory of computation that enables one to talk about programs as mathematical objects, analyzing them and proving things about them without reference to any particular machine. "Towards" concerns machines with state vectors and the changes wrought on them by storage, with or without side effects. Peter Landin's applicative expressions in whatever notation similarly served mathematical purposes, while the SECD machine occupied the middle ground between proofs and working code. One could say a lot about the mathematics of programs without touching the problem of assignment, but one could not write compilers for ALGOL or CPL, and a fortiori one could not write a semantically correct compiler-compiler.[29]

Hence, the work stimulated by McCarthy and Landin followed two lines of development during the late 1960s. One line probed the power and limits of recursion induction and other forms of reasoning about programs. Investigations in this area struck links with other approaches to verifying programs. The second line concerned itself with the mathematical problem of instantiating abstract systems in the concrete architecture of the stored-program computer. The two lines converged on a common mathematical foundation brought out by the work of Scott and subsequently pursued in the denotational semantics of Scott and Strachey and in a new algebraic approach proposed by Burstall and Landin.

Computer Science as Pure and Applied Algebra

From the start, McCarthy had looked to formal semantics not only to give definiteness to the specification of languages but also, and perhaps more importantly, to provide the basis for proving equivalences among programs and for demonstrating that a program will work as specified. While subsequent work considerably furthered the first goal and revealed the complexity of the programming languages being designed and implemented, it continued to speak a creole of formal logic and ALGOL. In the later 1960s, Burstall and Landin began to consider how the enterprise might be couched more directly in mathematical terms. As they wrote in a joint article published in 1969,

> Programming is essentially about certain "data structures" and functions between them. Algebra is essentially about certain "algebraic structures" and functions between them. Starting with such familiar algebraic structures as groups and rings algebraists have developed a wider notion of algebraic structure (or "algebra") which includes these as examples and also includes many of the entities which in the computer context are thought of as data structures.[30]

Working largely from Paul M. Cohn's *Universal Algebra* (1965), they expressed functions on data structures as homomorphisms of Ω-algebras, linking them via semigroups to automata theory and thereby extending to operations of higher arity on data structures the conclusion reached by J. Richard Büchi a few years earlier:

> If the definition of "finite automaton" is appropriately chosen, it turns out that all the basic concepts and results concerning structure and behavior of finite automata are in fact just special cases of the fundamental concepts (homomorphism, congruence relation, free algebra) and facts of abstract algebra. Automata theory is simply the theory of universal algebras (in the sense of Birkhoff) with unary operations, and with emphasis on finite algebras.[31]

By way of examples, Burstall and Landin set out a formulation of list-processing in terms of homomorphisms and a proof of the correctness of a simpler compiler similar to the one treated by McCarthy and Painter by means of recursion induction. Their approach can perhaps be summed up by one of their diagrams (see fig. 9).[32]

Category theory did not lie far beyond Ω-algebras and over the course of the 1970s became the language for talking about the typed λ-calculus,

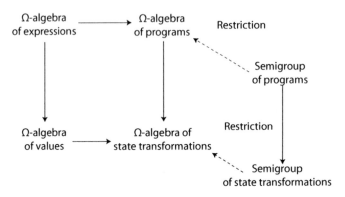

Figure 9 A formulation of list-processing in terms of homomorphisms and a proof of the correctness of a simpler compiler. Based on R. M. Burstall and P. J. Landin, "Programs and their proofs, an algebraic approach," *Machine Intelligence* 4 (1969), p. 32.

as functional programming in turn became in Burstall's terms a form of "electronic category theory."[33]

As automata theory, formal languages, and formal semantics took mathematical form at the turn of the 1970s, the applications of abstract algebra to computers and computation began to feed back into mathematics as new fields in themselves. In 1973 Cohn presented a lecture on "Algebra and Language Theory," which later appeared as chapter 11 of the second edition of his *Universal Algebra*. There, in presenting the main results of Schützenberger's and Chomsky's algebraic theory, he spoke of the "vigorous interaction [of mathematical models of languages] with some parts of noncommutative algebra, with benefit to 'mathematical linguist' and algebraist alike." A few years earlier, as Birkhoff was publishing his new textbook with Bartee in 1969, he also wrote an article reviewing the role of algebra in the development of computer science. There he spoke of what he had learned as a mathematician from the problems of minimization and optimization arising from the analysis and design of circuits and from similar problems posed by the optimization of error-correcting binary codes. Together,

> [these] two unsolved problems in binary algebra ... illustrate the fact that *genuine applications can suggest simple and natural but extremely difficult problems*, which are overlooked by pure theorists. Thus, while working for 30 years (1935–1965) on generalizing Boolean algebra to lattice theory, I regarded finite Boolean algebras as trivial because they could all be described up to isomorphism, and completely ignored the basic shortest form and optimal packing problems described above.[34]

Earlier in the article, Birkhoff had pointed to other ways in which the problems of computing are influencing algebra. To make the point, he compared the current situation with the Greek agenda of rationalizing geometry through constructions with ruler and compass (as analog computers).

> By considering such constructions and their optimization in depth, they were led to the existence of irrational numbers, and to the problems of constructing regular polygons, trisecting angles, duplicating cubes, and squaring circles. These problems, though of minor technological significance, profoundly influenced the development of number theory.
>
> I think that our understanding of the potentialities and limitations of algebraic symbol manipulation will be similarly deepened by attempts to solve problems of optimization and computational complexity arising from digital computing.

Birkhoff's judgment, rendered at just about the time that theoretical computer science was assigned its own main heading in *Mathematical Reviews*, points to just one way in which computer science was opening up a previously unknown realm of mathematics lying between the finite and the infinite, namely, the finite but intractably large. Computational complexity was another.

In 1974, acting in Scott's words as "the Euclid of automata theory," Samuel Eilenberg undertook to place automata and formal languages on a common mathematical foundation underlying the specific interests that had motivated them. "It appeared to me," he wrote in the preface of his intended four-volume *Automata, Languages, and Machines*,

> that the time is ripe to try and give the subject a coherent mathematical presentation that will bring out its intrinsic aesthetic qualities and bring to the surface many deep results which merit becoming part of mathematics, regardless of any external motivation.[35]

Yet, in becoming part of mathematics the results would retain the mark characteristic of their origins. All of Eilenberg's proofs were constructive in the sense of constituting algorithms. "A statement asserting that something exists is of no interest unless it is accompanied by an algorithm (i.e., an explicit or effective procedure) for producing this 'something.'" In addition, Eilenberg held back from the full generality to which abstract mathematicians usually aspired. Aiming at a "restructuring of the material along lines normally practiced in algebra," he sought to reinforce the original motivations rather than to eradicate them, arguing that both mathematics and computer science would benefit from his approach.

To the pure mathematician, I tried to reveal a body of new algebra, which, despite its external motivation (or perhaps because of it) contains methods and results that are deep and elegant. I believe that eventually some of them will be regarded as a standard part of algebra. To the computer scientist I tried to show a correct conceptual setting for many of the facts known to him (and some new ones). This should help him to obtain a better and sharper mathematical perspective on the theoretical aspects of his researches.

Coming from a member of Bourbaki, who insisted on the purity of mathematics, Eilenberg's statement is all the more striking in its recognition of the applied origins of "deep and elegant" mathematical results.

In short, in the formation of theoretical computer science as a mathematical discipline the traffic traveled both ways. Mathematics provided the structures on which a useful and deep theory of computation could be erected. In turn, theoretical computer science gave physical meaning to some of the most abstract, useless concepts of modern mathematics: semigroups, lattices, Ω-algebras, categories. In doing so, it motivated their further analysis as mathematical entities, bringing out unexpected properties and relationships among them. It is important for the historian to record this interaction before those entities are "naturalized" into the axiomatic presentation of the field, as if they had always been there or had appeared out of thin air.

The Structures of Computation and the Structures of the World

As noted earlier, in seeking a theory for the new electronic digital computer, John von Neumann took his cue from biology and neurophysiology. He thought of automata as artificial organisms and asked how they might be designed to behave like natural organisms. His program split into two lines of research: finite automata and growing automata as represented by cellular automata. By the late 1960s, the first field had expanded into a mathematically elegant theory of automata and formal languages resting firmly on algebraic foundations. At this point, the direction of inspiration reversed itself. In the late 1960s, Aristid Lindenmayer turned to the theory of formal languages for models of cellular development. Although he couched his analysis in terms of sequential machines, it was soon reformulated in terms of formal languages, in particular context-free grammars. His L-systems, which soon became a field of study themselves, were among the first applications of theoretical computer science and its tools to the biological sciences, a relationship reinforced by the resurgence in the 1980s of cellular automata as models of complex adaptive systems. During the 1990s theoretical chemist Walter

Fontana used the λ-calculus to model the self-organization and evolution of biological molecules. Most recently the λ-calculus, initially designed by Robin Milner by analogy to the λ-calculus for defining the syntax and operational semantics of concurrent processes, has become the basis for analyzing the dynamics of the living cell. As the introduction to a recent volume describes the relationship,

> The challenge of the 21st century will be to understand how these [macro-molecular] components [of cellular systems] integrate into complex systems and the function and evolution of these systems, thus scaling up from molecular biology to systems biology. By combining experimental data with advanced formal theories from computer science, "the formal language for biological systems" to specify dynamic models of interacting molecular entities would be essential for: (i) understanding the normal behaviour of cellular processes, and how changes may affect the processes and cause disease . . . (ii) providing predictability and flexibility to academic, pharmaceutical, biotechnology and medical researchers studying gene or protein functions.[36]

Here the artifact as formal (mathematical) system has become deeply embedded in the natural world, and it is not clear how one would go about re-establishing traditional epistemological boundaries among the elements of our understanding.

Over the past three decades, computational models have become an essential tool of scientific inquiry. They provide our only access, both empirically and intellectually, to the behavior of the complex non-linear systems that constitute the natural world, especially the living world. As Paul Humphrey notes concerning the massive quantities of data that only a computer can handle,

> Technological enhancements of our native cognitive abilities are required to process this information, and have become a routine part of scientific life. In all of the cases mentioned thus far, the common feature has been a significant shift of emphasis in the scientific enterprise away from humans because of enhancements without which modern science would be impossible . . . [I]n extending ourselves, scientific epistemology is no longer human epistemology.[37]

In other words, our understanding of nature has come to depend on our understanding of the artifact that implements those models, an artifact that we consider to be essentially mathematical in nature: it is the computation that counts, not the computer that carries it out. Thus, our epistemology retains a human aspect to the extent that mathematics affords us conceptual access to the computational processes that enact our representations of the world. And that remains an open problem.

In 1962, looking ahead toward a mathematical theory of computation and back to the standard of a mathematical science set by Newton, John McCarthy sought to place computing in the scientific mainstream:

> It is reasonable to hope that the relationship between computation and mathematical logic will be as fruitful in the next century as that between analysis and physics in the last. The development of this relationship demands a concern for both applications and for mathematical elegance.[38]

The relationship between analysis and physics in the nineteenth century had borne more than theoretical fruit. It had been the basis of new technologies that transformed society and indeed nature. Implicit in McCarthy's exhortation, then, was a mathematical science of computation that worked in the world in the way that mathematical physics had done. McCarthy had in mind, of course, artificial intelligence and its practical applications.

During the 1970s the new sciences of complexity changed mathematical physics. In 1984 Stephen Wolfram similarly looked to past models for a mathematical science of a decidedly non-Newtonian world. In laying out the computational theory of cellular automata, he noted that:

> Computation and formal language theory may in general be expected to play a role in the theory of non-equilibrium and self-organizing systems analogous to the role of information theory in conventional statistical mechanics.[39]

Table 1, derived from his article, reflects what he had in mind.

But with the power of computational modeling came a new challenge to mathematics. Traditional models enhance understanding by mapping the operative parameters of a phenomenon into the elements of an analyzable mathematical structure, e.g., the mass, position, and momentum of a planet related by a differential equation. Computational models, in particular computer simulations, render both the mapping and the dynamics of the structure opaque to analysis. While the program may help to discern the relation between the structure of the model and that of the phenomenon, the process determined by that program remains all but impenetrable to our understanding. As Christopher Langton observed,

> We need to separate the notion of a formal specification of a machine—that is, a specification of the logical structure of the machine—from the notion of a formal specification of a machine's behavior—that is, a specification of the sequence of transitions that the machine will undergo. In general, we cannot derive behaviours from structure, nor can we derive structure from behaviours.[40]

Table 1. Cellular Automata and Formal Languages

	Qualitative characterizations of complexity	Patterns generated by evolution from initial configuration	Effect of small changes in initial configurations	"Information" of initial state propagates	Formal language representing limit set
1	Tends toward a spatially homogeneous state	Pattern disappears with time	No changes in final state	Finite distance	Regular
2	Yields a sequence of simple stable or periodic structures	Pattern evolves to a fixed finite size	Changes only in a region of finite size	Finite distance	Regular
3	Exhibits chaotic aperiodic behavior	Pattern grows indefinitely at a fixed rate	Changes over a region of ever-increasing size	Infinite distance at fixed positive speed	Context sensitive?
4	Yields complicated localized structures, some propagating	Pattern grows and contracts with time	Irregular changes	Infinite distance	Universal

Source: Stephen Wolfram, "Computation Theory of Cellular Automata," in Communications in Mathematical Physics 96 (1984).

In the concluding chapter of *Hidden Order: How Adaptation Builds Complexity*, Holland makes clear what is lost thereby. Looking Toward Theory and the general principles that will deepen our understanding of *all* complex adaptive systems [*cas*], he insists as a point of departure that:

> Mathematics is our sine qua non on this part of the journey. Fortunately, we need not delve into the details to describe the form of the mathematics and what it can contribute; the details will probably change anyhow, as we close in on our destination. Mathematics has a critical role because it along enables us to formulate *rigorous* generalizations, or principles. Neither physical experiments nor computer-based experiments, on their own, can provide such generalizations. Physical experiments usually are limited to supplying input and constraints for rigorous models, because the experiments themselves are rarely described in a language that permits deductive exploration. Computer-based experiments have rigorous descriptions, but they deal only in specifics. A well-designed mathematical model, on the other hand, generalizes the particulars revealed by physical experiments, computer-based models, and interdisciplinary comparisons. Furthermore, the tools of mathematics provide rigorous derivations and predictions applicable to all *cas*. Only mathematics can take us the full distance.[41]

In the absence of mathematical structures that allow abstraction and generalization, computational models do not say much. Nor do they function as models traditionally have done in providing an understanding of nature on the basis of which we can test our knowledge by making things happen in the world.

In pointing to the need for a mathematics of complex adaptive systems, Holland was expressing a need as yet unmet. The mathematics in question "[will have to] depart from traditional approaches to emphasize persistent features of the far-from-equilibrium evolutionary trajectories generated by recombination." His sketch of the specific form the mathematics might take suggests that it will depart from traditional approaches along branches rather than across chasms, and that it will be algebraic. But it has not yet been created. If and when it does emerge, it is likely to confirm Schützenberger's principle that "it is again a characteristic of the axiomatic method that it is at the same time always a new chapter in mathematics."[42]

CHAPTER THIRTEEN

Extracts from *Software as Science—Science as Software*

Editorial note: This version of the material Mahoney published in "Computers and Mathematics" and "The Structures of Computation and the Mathematical Structure of Nature" was presented at a conference devoted to "mapping" issues in the history of software. Delivering one of five keynote addresses, Mahoney was responsible for exploring software as science, a topic his interest in modeling inspired him to complement with its inversion. Several extracts are included here of material unique to this presentation of his work.

Extract 1: Agendas and Institutions in Computer Science

Software should be of great interest to historians of science. That may seem strange, given that it is of such recent origin. Software is no older than the modern electronic computer and the activity of writing programs for it. It is still experiencing growing pains. Yet, over the past fifty years, it has become the subject of its own thriving science and a ubiquitous medium for pursuing other sciences. In both instances software represents a new kind of science. It is what Herbert Simon calls a "science of the artificial." There is nothing natural about software or any science of software. Programs exist only because we write them, we write them only because we have built computers on which to run them, and the programs we write ultimately reflect the structures of those computers. Computers are artifacts, programs are artifacts, and models of the world created by programs are artifacts. Hence, any science about any of these must be a science of a world of our own making rather than of a world presented to us by nature. What makes it both challenging and intriguing is that those two worlds meet in the physical computer, which enacts a program in the world. Their encounter has posed new and difficult epistemological questions concerning what we can know both about the workings of the models and about the relation of the models to the phenomena they purport to represent or simulate. Answers to those questions would seem to depend, at least in part, on understanding programs as dynamic systems.[1]

Because software as science is both new and artificial, it brings to the fore questions of when and how and who. It took some time before programs and programming became subjects of inquiry in themselves. Once they did, it was not clear what one wanted to know about the programs or the activity of writing them, or indeed could know about these subjects. Debate seems to have been particularly lively in the late 1960s. Most practitioners viewed the subject as inherently mathematical. Yet, Marvin Minsky decried excessive formalism, pointing to the "defeatism" of theorems about the limits of computability and to the inadequacy of formal systems to provide an explanatory account of what computers could actually do. Allan Newell and Herbert Simon insisted even more strongly on treating computer science as an empirical discipline. It is the study of the computer as a dynamic physical device and what programmers are capable, intentionally or not, of making it do. Donald Knuth insisted on the craft nature of programming, characterizing it as an "art." In what began as a "light-hearted attempt to stir up some controversy regarding the nature of computer science," Peter Wegner tried in "Three Computer Cultures" to differentiate the concerns of the computer science from those of both the mathematician and the engineer. While George Forsythe offered counsel on "What to Do until the Computer Scientist Comes," John Pierce ended the keynote address cited above with the hope that "computer scientists, whatever they are, get organized effectively, and I wish them good luck."[2]

So the history with which we are concerned begins with the question of who created the science(s) of software, when, where, why, and how? That is, who thought it necessary or desirable to place programs and programming on some sort of scientific foundation? What was such a foundation meant to accomplish? To what questions would it provide answers? What theoretical and practical benefits did it promise? What sort of science did its creators envision? That is, what established scientific disciplines did they take as models and resources for their new enterprise? How did these aspirations shape the science(s) that emerged, and how did the development of the science(s) reshape the aspirations? As the science(s) developed, how did it (or they) interact with other sciences, especially those that looked to the computer as a tool and then as a medium of investigation? What did those sciences contribute to the science of software and what in turn did they take from it? What about other disciplines not generally considered scientific?

These questions clearly intersect with those of the other areas on our program. The mathematical verification of programs as a warrant of reliability lies at the root of formal semantics. Moreover, if one views engineering as applied science, then one faces the question of what science it is that software engineering applies. Conversely, one may ask what role the theory that has emerged has played in the practice of programming, especially programming in the large, and how that role fits with the status accorded to theory (and to those who pursue it) in the profession at large. The answers to those questions impinge in evident ways on the nature and organization of software as a labor process.

The science mainly in question here is mathematics, the relation of which to computing has evolved dynamically over the past half century. As a physical device, the modern computer was built for mathematicians to carry out numerical calculations, especially for problems which could not be solved analytically. As a theoretical concept, the modern computer was designed by mathematical logicians to understand the nature of computability and the limits of what can be known by it. As a dynamic computational system, the modern computer has posed new mathematical problems and opened new fields of mathematical research. At the same time, the computer has proved elusive, as central concerns of programming remain beyond the effective reach of mathematics and thus again raise the question of what software as science has to do with software as engineering or as reliable artifact.

No differently from any other science, software as science involves more than a body of knowledge and practice. It means communities of practitioners recognized as possessing that knowledge and charged with extending and disseminating it. The science in question is what they know and do in common. Taking this approach allows for the science(s) of programs and programming to take different forms among different groups of practitioners. That is, it allows for different answers at different times in different places to the question "what is software and what may be said scientifically about it?," or indeed, "what needs to be said scientifically at all about software?" To the extent that a consensus has emerged, it requires an explanation, and historians of science have found that the explanation is likely to be as much social as intellectual.

Over the past fifty years, computer scientists have grown from a handful of people to an extensive network of practitioners in industry, academia, and private practice. They occupy positions of prominence in colleges and universities; indeed, together with molecular biology (with which they have intellectual ties), they constitute the fastest-growing sector of academia. Generously funded by industry and government, they have professional associations (ACM, IEEE Computer Society, BCS, etc.), journals, monographs, textbooks, and an elaborate reward structure. Much of this growth rests on a claim to be pursuing a scientific enterprise, even as practitioners have debated among themselves just how scientific it is or should be. How practitioners achieved recognition of that claim is an integral part of the history of software as science.[3]

Theoretical Computer Science

Elsewhere I have suggested that the practice of a discipline can be fruitfully approached through the notion of "agenda."* That tools embody agendas has particular importance for new sciences. For a new science means a new agenda, and tracing the emergence of a new science means showing how a group of practitioners coalesced around a common agenda different from other agendas in which they had been engaged. What questions or problems drew them to the computer? What tools did they bring with them and how did they apply those tools? How did their involvement shape the emerging agenda of the new field?

That brings me to what is generally considered the scientific basis of software, namely, theoretical computer science. It took shape between 1955 and 1975 as practitioners from a variety of fields converged on a small set of related agendas that came to constitute the core of the field: automata and formal languages, computational complexity, and formal semantics. None of those agendas had existed before 1955. By the early 1970s their status as constituents of an autonomous discipline was marked by a main heading in *Mathematical Reviews*, by a growing number of dedicated textbooks, and by the establishment of curricula at both the undergraduate and graduate levels. Perhaps even more strikingly, by the mid-1970s theoretical computer science had begun to gain recognition as a field of mathematics in its own right and to serve as resource for other sciences, most notably theoretical biology.[4]

My charge is not to provide a history of that development but to suggest what such a history might look like and how it might be most productively pursued, in short, to offer an agenda for the history of software viewed as science. So let me restrict my account to the following diagrams

---

* Editorial note: At this point Mahoney reproduced two paragraphs on agendas from "Computer Science: The Search for a Mathematical Theory." They have been removed from this version to avoid duplication.

which encapsulate my own efforts to trace the emergence of the agendas of theoretical computer sciences from the intersection and interaction of a variety of agendas in fields ranging from electrical engineering to linguistics.* The schemes suggest a number of lines of fruitful inquiry.[5]

To begin with, it seems clear that theoretical computer science can be viewed from a number of disciplinary perspectives. Indeed, its formation can be understood only from those perspectives. Computing had no science of its own at the start. Mathematical logic had established what computers could not do, even with endless resources of time and space. Switching theory showed how to analyze and synthesize circuits for basic operations. But no science accounted for what finite machines with finite, random access memories could do or how they did it. That science had to be created, and its creation depended heavily on what was going on in other fields at the time, most notably linguistics. Before the science of computing began to accumulate a history of its own, it was heir to several different histories. Understanding its subsequent development may well involve keeping those histories in mind and looking for their continuing influence.

At each point of convergence the nascent field acquired a set of tools from an antecedent discipline. One may ask what those tools were originally developed to accomplish, how their application to computing contributed to shaping the new subject, to what extent the application in turn reshaped the tools or redefined their status in the parent discipline. An example is the new mathematical interest acquired by finite Boolean algebras as a result of their application to questions of the minimization and optimization of sequential circuits.[6]

Indeed, one of the uncanny aspects of the development of theoretical computer science has been the way it has given practical meaning to the most abstract mathematical structures: semigroups, lattices, categories. None of these was created with computers in mind, and in each case it is not hard to find statements by mathematicians of the time insisting on their uselessness even to mathematics. Each is fundamental to modern computer science, which has arguably created the notion of "applied algebra," even to the point that one recent book offers "category theory for the working computer scientist." Though not originally a mathematical construct, the lambda calculus has similarly moved from theoretical structure to practical tool (especially once Scott provided a mathematical model in continuous lattices) and indeed recently has begun to move out from computing per se into the area of theoretical biology.[7]

---

* Editorial note: See chapter 12, "The Structures of Computation and the Mathematical Structure of Nature," for these diagrams (figures 6–9).

The diagrams indicate, however sketchily, that various parts of the agenda took shape initially in different places. For example, the identification of formal power series, pushdown automata, and context-free languages brought together at MIT agendas ranging from the algebraic coding theory of Marcel Schützenberger in Paris to the work on sequential formula translation of Fritz Bauer and Klaus Samelson in Munich, which in turn drew on Heinz Rutishauser's early efforts at automatic programming. To take another example, the notion of using the lambda calculus as the basis for formal semantics seems clearly to have originated with John McCarthy, who needed it as a means of abstracting functions for his work on mechanical theorem-proving and on commonsense reasoning by computers. Yet, it seems equally clear from activities surrounding the ALGOL meetings that others besides McCarthy were familiar with the lambda calculus and were exploring its use as a vehicle for defining the semantics of the new language. Indeed, the lambda calculus and formal semantics quite quickly crossed the Atlantic in the early 1960s, settling in primarily with Peter Landin and Christopher Strachey at Cambridge but then spreading to Vienna and Amsterdam, where Dana Scott's seminal collaboration with Jaco de Bakker took place. While McCarthy's work spoke to an agenda already under way elsewhere, transcripts of a Working Conference on Mechanical Language Structures held in Princeton in 1963 suggest that it received a cooler reception closer to home, where more pragmatic concerns dominated.[8]

These are just two examples, I suspect, of how agendas at first reflect local interests and ways of doing things. Different groups of people constitute different mixes of scientific training, taste, and aspirations, reflecting in many cases their differing cultural and institutional backgrounds. In addition to questions of how a local group coalesces around a common project, there is the larger issue of how that project then moves onto the agenda of the profession as a whole. Of great interest among historians of science over the past decade has been a question of how practices travel. Studies have revealed the particular importance of individuals moving from one place to another, learning and conveying by collaboration and example results and techniques that have not yet reached print.[9]

Research and Training

One measure of the importance of an agenda is the resources allocated to it by the community of practitioners, usually acting as agents for the government or industry. Norberg and O'Neill's study of DARPA's IPTO and the as yet unpublished study of National Science Foundation's Office of Computing Activity by Aspray, Williams, and Goldstein offer glimpses into the interactive process by which government agencies and the research community shape the agenda of the discipline. As in so many other instances, Dick Hamming offers historians a valuable perspective on what was at stake. In his Turing Lecture of 1968, at a time when the nature of the field seemed uncertain, he warned his audience:

> In the face of this difficulty [of defining "computer science"] many people, including myself at times, feel that we should ignore the discussion and get on with *doing* it. But as George Forsythe points out so well in a recent article, it *does* matter what people in Washington D.C. think computer science is. According to him, they tend to feel that it is a part of applied mathematics and therefore turn to the mathematicians for advice in the granting of funds. And it is not greatly different elsewhere; in both industry and the universities you can often still see traces of where computing first started, whether in electrical engineering, physics, mathematics, or even business. Evidently the picture which people have of a subject can significantly affect its subsequent development. Therefore, although we cannot hope to settle the question definitively, we need frequently to examine and to air our views on what our subject is and should become.[10]

Research funding is a matter of more than just money. Until a field gains autonomy over its own agenda, its development depends on what other disciplines think its practitioners should be doing.

Proposals and requests for proposals (RFPs) provide valuable insight into the articulation of agendas. They aim at enlisting support and hence must tie the proposers' aims to those of their reviewers and their reviewers' institutions. Moreover, they capture the proposers' thinking before the work has been carried out and thus offer a chance of comparing shifts in direction as questions are answered, sometimes in unanticipated ways. It is revealing, for example, to see how Minsky, McCarthy, Shannon, and Rochester viewed the agenda they called "artificial intelligence" in proposing their famous summer workshop in 1956. McCarthy has subsequently insisted on the value of such documents in establishing the aims and methods of scientific research.[11]

Viewing a science in terms of its evolving agenda means, among other things, seeing how the agenda is communicated to the next generation of practitioners. In explaining what I mean by "agenda," I said that one becomes a practitioner by learning what is to be done, i.e., by learning what the questions or problems of the field are, how they are tackled and resolved, and what constitutes a solution. Kuhn's notion of "paradigm" fits well here, especially as he subsequently clarified it through the concept of a "disciplinary matrix." Science is conveyed by exemplars, by models of problem solving. Students start with what is best established and most familiar to practitioners, and then move from there onto rougher terrain until they come to the edges of known territory. In the sciences in particular, that does not mean that students must recapitulate the entire history of the discipline. On the contrary, what makes certain solutions paradigmatic is precisely the way in which they encompass and give structure to a range of problems, transforming their hard-won solutions into corollaries.[12]

That is what makes textbooks and curricula an important resource for tracing the emergence and development of a discipline. They reflect its agenda not at the frontiers of research but at the starting point for reaching those frontiers. They are statements about what current practitioners at a particular time think students must know to become the next generation of practitioners. Hammering out a curriculum can be a harrowing experience for participants precisely because it means reaching agreement on what the subject is about: what is central and what peripheral, what must everyone know and what can be an elective, in what order are these things to be learned? To the historian, the process of hammering out is as important as, or perhaps even more important than the end result. We have the published versions of a succession of ACM curricula in computer science and responses to them. I hope we also have the minutes of the meetings of the committees that wrote them, not to say copies of the exchanges that went on between meetings.[13]

What the profession as a whole was trying to accomplish was happening in colleges and universities, as computer scientists sought to define a place for themselves in their institutions and to justify their recognition as distinct academic units on a par with those already established. The volume on *University Education in Computing Science* edited by Aaron Finerman in 1968 provides a good survey of the range of thinking on the matter at that crucial time. The local strategies of practitioners, in particular the alliances they forged with other disciplines, should also prove revealing. Anniversaries and retirements of founders have provided largely celebratory accounts for departments at Purdue, Cornell, MIT, and elsewhere, but no one has yet undertaken a critical, documented analysis of how the new science of computing established itself at a university.[14]

Extract 2: Mathematics and Software Engineering

Editorial note: Mahoney summarized John McCarthy's hope that the relationship between computation and mathematical logic would prove as fruitful as the historical relationship of physics and mathematics, using the same pair of quotations found in several of his other papers but putting them in the context of broad faith in the power of basic science during the 1950s and '60s. He then returned briefly to the 1969 Rome conference on software engineering as another place where mathematical work on computer science was challenged to show its relevance.

Ten years later, the Computer Science and Engineering Research Study (COSERS) took stock of the field and its current directions of research and published the results under the title *What Can Be Automated?* The committee on theoretical computer science argued forcefully that a process of abstraction was necessary to understand the complex systems constructed on computers and that the abstraction "must rest on a mathematical basis." Defining theoretical computer science as "the field concerned with fundamental mathematical questions about computers, programs, algorithms, and information processing systems in general," the committee acknowledged that those questions tended to follow developments in technology and its application, and hence to aim at a moving target—strange behavior for mathematical objects.[15]

Nonetheless, the committee could identify several broad issues of continuing concern, which were being addressed in the areas of computational complexity, data structures and search algorithms, language and automata theory, the logic of computer programming, and mathematical semantics. In each of these areas, it could point to substantial achievements in bringing some form of mathematics to bear on the central questions of computing. Yet, in the summaries at the end of each section, they repeatedly echoed Christopher Strachey's lament. For all the depth of results in computational complexity, "the complexity of most computational tasks we are familiar with—such as sorting, multiplying integers or matrices, or finding shortest paths—is still unknown." Despite the close ties between mathematics and language theory, "by and large, the more mathematical aspects of language theory have not been applied in practice. Their greatest potential service is probably pedagogic, in codifying and given clear economical form to key ideas for handling formal languages." Efforts to bring mathematical rigor to programming quickly reach a level of complexity that makes the techniques of verification subject to the very concerns that prompted their development. Mathematical semantics could show "precisely why [a] nasty surprise can arise from a seemingly well-designed programming language," but not how to eliminate the problems from the outset. As a design tool, mathematical semantics was still far from the goal of correcting the anomalies that gave rise to errors in real programming languages. If computers and programs were "inherently mathematical objects," the mathematics of the computers and programs of real practical concern had so far proved elusive.

Five years later, C. A. R. Hoare echoed the committee's expression of belief and admission of fact. In a postponed Inaugural Lecture as Professor of Computation at Oxford in 1985 (he had been appointed in 1976), Hoare declared,

Our principles may be summarized under four headings.

(1) Computers are mathematical machines. Every aspect of their behavior can be defined with mathematical precision, and every detail can be deduced from this definition with mathematical certainty by the laws of pure logic.

(2) Computer programs are mathematical expressions. They describe with unprecedented precision and in every minutest detail the behaviour, intended or unintended, of the computer on which they are executed.

(3) A programming language is a mathematical theory. It includes concepts, notations, definitions, axioms and theorems, which help a programmer to develop a program which meets its specification, and to prove that it does so.

(4) Programming is a mathematical activity. Like other branches of applied mathematics and engineering, its successful practice requires determined and meticulous application of traditional methods of mathematical understanding, calculation and proof.

These are general philosophical and moral principles, and I hold them to be self-evident—which is just as well, because all the actual evidence is against them. Nothing is really as I have described it, neither computers nor programs nor programming languages nor even programmers.

In the first three cases, sheer size and complexity stood in the way of mathematical understanding. In the case of programmers, "ignorance or even fear of mathematics" blocked many, while those trained in mathematics did not apply it.[16]

What should interest the historian of science here is a continuing dissonance between the premises of theoretical computer science and the experience of programming. It constitutes a prime example of how modern technoscience confounds traditional categories of theory and practice. In principle, the computer should be accessible to mathematics. In practice, it is not.[17]

There are two features of the situation which have implications for how we might approach the history of software. First, as COSERS itself observed, "Even though all the levels of the hierarchy which computer systems can be interpreted as algorithms, the *study of algorithms* and the *phenomena related to computers* are not coextensive, since there are important organizational, policy, and nondeterministic aspects of computing that do not fit the algorithmic mold." The observation raises the questions of what mold those aspects do fit, that is to say, what science, if any, encompasses the phenomena not covered by algorithms. The second feature is the complexity of computer systems that seems to place even their algorithmic aspects beyond the reach of mathematics. The first feature has implications for software as engineering, the second for science as software.[18]

Viewing Hoare's principles as ideals to be pursued by ever more rigorous methods risks misleading both the historian and the software engineer. To see why, consider the following variation on the traditional "waterfall" model of software development (fig. 5). Relatively few errors now occur in the stages in the bottom half of the scheme. That is not surprising, given that they are the aspects of computing best understood mathematically, and that understanding has been translated into such practical tools as diagnostic compilers for high-level programming languages.

However, as Fred Brooks pointed out in "No Silver Bullet," the problems thus addressed were accidental, rather than essential, to the task of designing large, complex computer systems. At the top of the scheme, the situation is different. That is where the bulk of the crucial errors have been made, and that is where software engineering has focused its attention since the 1970s. But that is also where the science of software moves away from the computer into the wider world and interacts with the sciences (if they exist) pertinent to the systems to be modeled computationally. There it becomes a question of how to express those sciences computationally and of how to evaluate the fit between the target system and the computational model. But that is a question that software engineers share with scientists who have turned to the computer to take them into realms that are accessible neither to experiment nor to analytical mathematics. Intellectually, professionally, and historically, it links software as science to science as software.[19]

Extract 3: Software as Science

Editorial note: Mahoney included this section in the published version of the paper as a response to issues arising in discussion at the conference.

In common English usage, "software" is a mass term for programs, for what computers process, as opposed to the machines themselves. When the term first arose in the late 1950s, it was used as the antonym of "hardware." As the *Oxford English Dictionary* defines it, noting its formation on the model of "hardware," software is "the programs and procedures required to enable a computer to perform a specific task, as opposed to the physical components of the system." By the mid-1960s, the term had taken on a more specific sense of systems software, what people use to construct and run programs. That is what John Tukey had in mind when he introduced the term in 1958. But the term retained its broader meaning; the "software houses" that sprang up in the mid-1960s were producing applications rather than systems.[20]

In my essay "software" means simply programs and the activity of writing them, programming. In speaking then of "software as science" I do not mean to assert that programs in and of themselves constitute a science, or that the writing of them is an inherently scientific activity. Clearly, neither is the case. As I pointed out at the start, programs are artifacts—literally, things crafted—and they are no more inherently scientific than any other made object in the world. However, just as other sciences have arisen from the investigation of artifacts, e.g., thermodynamics from the steam engine, so one may ask about a science of programs and programming. Or rather, one may look for efforts by practitioners of computing to make programs and programming the subjects of a scientific inquiry, to place them on a scientific foundation. That is how I construe "software as science" for the purposes of historical inquiry.

To judge from the discussion, people in computing evidently disagree about whether such a science is possible, desirable, or relevant. That is not a matter for historians to decide, nor do historians require consensus on the matter among computer people; indeed, the continuing disagreement is of greater interest than consensus. For historians it is enough that from the mid-1950s people of reputation in computing have believed that a science of software is desirable and feasible and have set forth what they take that science to be. Most of them have looked to mathematics as the foundation. Tony Hoare represents perhaps the extreme in his insistence on the inherently mathematical nature of programs and programming, but he hardly stands alone out there. Others have taken a more empirical approach, believing that mathematics is not adequate to explain what computers can do, especially when we have not told them (or do not believe we have told them) to do it. But, as envisioned by Herbert Simon, the empirical science of computational processes would be no less scientific for being empirical and, indeed, a science of the artificial.

To justify a history of the science of software, it would seem enough to point to the two-volume *Handbook of Theoretical Computer Science*, of which neither the contents nor even the constituent subjects existed in 1950. The extensive bibliographies accompanying each of the thirty-seven chapters testify to the immense intellectual effort and hence social investment that the *Handbook* is meant to codify. My sketch of the development of just two of those constituents, formal languages and formal semantics, aimed at suggesting how historians might go about tracing the origins and growth of the field and determining how it attracted the investment. I am fairly confident that the notion of agendas and their convergence on the computer will go a long way in explaining the emergence of such subjects as computational complexity, databases, computational geometry, and parallel computing. Whether or not it does, the *Handbook* evidently purports to constitute a science of software and to base that science for the most part on mathematics. The job of the historian is not to question whether the *Handbook* should exist but to explain how it came about.[21]

The *Handbook* speaks to another point raised in the discussion. Several in the audience suggested that in speaking of "software as science" I was laboring under a misapprehension rooted in the English use of "computer science" to denote a subject other languages refer to as "informatics." Perhaps the name was leading me to look for science where there was none. Considering that slightly more than half the authors of the *Handbook* are Europeans engaged in informatics, the objection is puzzling. It is all the more so, since I need only turn to my bookshelf to find a volume, *Theoretische Informatik—kurzgefaßt* by Uwe Schöning, the contents of which cover roughly the same ground as, say, *Computability, Complexity, and Languages: Fundamentals of Theoretical Computer Science* by Martin Davis, Ron Sigal, and Elaine J. Weyuker. Turning to Schöning's institutional homepage, the Fakultät für Informatik at the University of Ulm, I see *Theoretische Informatik I/II* as part of the *Grundstudium*, side by side with *Technische Informatik I/II*, laying the foundation for further study. The rest of the curriculum does not look substantially different from what is taught in any American department of computer science. So I must wonder, as Shakespeare's Juliet once did, "What's in a name?" Here too the rose smells the same.[22]

Extract 4: Are There Laws of Software?

Editorial note: In "The Structures of Computation and the Mathematical Structure of Nature" Mahoney mentions the reception computer scientists gave to his original presentations of "Software of Science—Science as Software" and defends his characterization of software as a potential object of scientific inquiry. His presentation of the argument in the published version of "Software as Science" included further discussion of parallels between laws governing software and those described in physics.

These two features [of mechanical philosophy: laboratory experimentation and mathematical certainty] come together in thermodynamics, the mathematical laws of which originated in the steam engine, abstracted by Carnot to the concept of a heat engine. The first and second laws in effect say that we can't build a perpetual-motion machine of the first or second kind. They are negative laws that set limits, without saying much about what can be achieved within those limits. Shannon's mathematical theory of communication similarly built a science out of artifacts, in this case various communications technologies. The theory sets a limit on channel capacity and specifies the trade-off between redundancy and accuracy. The fruitful interaction between thermodynamics and information theory in explaining physical phenomena, as in the work of Stephen Hawking, seems to work against any effort at distinguishing the natural from the artificial or the mathematical from the physical.

So too with the theory of computation. Does it have laws? Surely Turing's halting theorem sets a limit on what can be computed by demonstrating what cannot be. Showing that a problem is equivalent to the halting problem relegates it to the realm of the incomputable. Similarly, the theory of computational complexity establishes through a variety of models of limited Turing machines what resources are required to compute classes of problems and the trade-offs between time and space involved in doing so. Are these scientific results about nature? Well, there's a body of literature that says yes. It is a tenet of the new computational sciences that nature can't compute any better than a Turing machine. Or rather, anything nature can do, a computer can do too, given enough time and memory.[23]

Are these laws of software? In discussing the implications of software as science for software as engineering, I addressed the limits of theoretical computer science in addressing the software development process. That process begins with the translation of a portion of the real world into the first of a series of computational models which culminate in a program running on a specific machine. Verification of the result involves two different issues: the goodness of the computational model with respect to the world it is supposed to model and the accuracy of the translation of that model into the instruction set of the computer on which it is running. Whether the model itself is adequate is ultimately not a question of software but of the developer's understanding of the world. It may be a scientific question, but the science involved is not about software or computers. The science of software as I have construed it pertains to the second issue. It begins where the model becomes a program: how, and to what extent, can we assure ourselves that the program is doing what we have written it to do? How that question has been formulated and addressed is the subject of the history of software as science, or at least that is how I construed my charge.

Éloge

Michael Sean Mahoney, 1939–2008

Jed Z. Buchwald
D. Graham Burnett

PERHAPS THE CLEAREST testimony to the scholarly range and depth of Princeton's now lamented Michael S. Mahoney lies in the dismay of his colleagues in the last few years, as they contemplated his imminent retirement. How to maintain coverage of his fields? Fretting over this question, the program in history of science that he did so much to build recently found itself sketching a five-year plan that involved replacing him with no fewer than four new appointments: a historian of mathematics with the ability to handle the course on Greek antiquity, a historian of the core problems of the scientific revolution, a historian of technology who could cover the nineteenth-century U.S. and Britain, and finally, a historian of the computer-and-media revolution. In his passing we have lost a small department.

Best known for his exacting *The Mathematical Career of Pierre de Fermat, 1601–1665* (1973, revised 1994), Mahoney also authored a valuable translation of Descartes's *Le Monde* (1979) and saw a suite of his essential essays on seventeenth-century mathematics several times published in book form in Japanese (under a title that would be Englished as *Mathematics in History*, originally printed 1982, revised 2007). At the time of his death the better part of two manuscripts lay in the drawers of his book-filled office: a study of the mathematical thought of Christiaan Huygens, and a long-awaited volume on the history of computing and software engineering. It is to be hoped that both these works will yet find their way to print.

For many years a scroll of old-fashioned tractor-feed computer printout hung over his office door (on the inside, where he could read it from his desk) bearing in foot-high letters, birthday-banner style, the chastening motto of that towering figure of modern mathematics, Carl Friedrich Gauss: *Pauca sed Matura*, which Mahoney's Latinists (he started his career as a Medievalist, and continued to prefer to read Newton's *Principia* in the original) knew was generally translated "few, but ripe," as in, "not a ton of stuff, but all of it very good." One sensed that it hung as a banner strung between scholarly pride and quiet self-mortification.

Though of the latter there was little need. Over the course of a forty-year career Mahoney published more than sixty articles and book chapters, many scores of book reviews, and lectured on the history of science and technology from Tokyo and Beijing to Berlin and London, from Athens and Jerusalem to schoolteachers in suburban New Jersey. His intellectual reach and pedagogical generosity were fabled, and more than one graduate student can recall his improvisational lunchtime pencil-on-a-napkin reconstructions of the besetting geometrical paradoxes of the Pythagoreans, just as two full generations of undergraduates retain gulp-inducing memories of his drop-this-class-if-you-don't-want-to-sweat opening lecture in History 291, his class on the scientific revolution, in which, using a computer program of his own confection, he demonstrated the intricacies of the Ptolemaic cosmography, complete with dizzying loops upon loops of epicycles, eccentric deferents, and a speckling of equant points. The lesson was two-fold: first, history of science was going to require that everyone gird up for problem sets; second, superseded explanations of natural phenomena might indeed have been wrong, but they were finely wrought things, rigorous and, in context, generally very effective. Upshot? Historical understanding of the sciences would require both hard work and the sympathies of the imagination. If he had put his own motto up on the wall, it might have been that.

Born in New York City on the thirtieth of June 1939, Mahoney graduated *magna cum laude* from Harvard College in the class of 1960 with a double concentration in history and science. Mathematics, however, was his first love, and with the support of a German Foreign Service Fellowship he spent the next two years in Munich, studying with the celebrated historian of mathematics, Kurt Vogel, whose late-career appetite for the computational quirks of Babylonians and Egyptian calendrical calculations rubbed off on the young American and left a shimmer. Returning to the United States to pursue a doctorate in the history of science in Charles Gillispie's newly created program at Princeton, Mahoney moved away from the archaeo-philological sifting of antiquity and took up the role of mathematics in the rise of the new sciences, a topic he would never entirely put down. His gifts as a scholar and teacher quickly recognized, Mahoney had the good fortune to be converted from graduate student to member of the faculty at Princeton, acceding to full professorship in 1980.

The Princeton program in its early years was housed in a small building on Washington Road, where Mahoney, Ted Brown, and Tom Kuhn had their offices, and where graduate seminars were held. Mahoney worked with Kuhn as a "preceptor" in Kuhn's course of lectures on the entire history of science, and it was during those early years that Mahoney inaugurated his own courses. He did not dabble lightly in early science, and neither were his students allowed to. Mahoney had, for example, translated many of Huygens's papers on the pendulum and on colliding bodies, and his students were expected to work line-by-line through the material, to take it apart and put it back together in ways that revealed the essential go of the arguments. He expected hard work and was masterful at guiding discussion in ways that permitted students to show what they could do. Both of us studied with Mahoney—Buchwald in the late 1960s, Burnett in the early 1990s—and throughout all those years Mahoney's respect for, and interest in, his students never changed.

Mahoney's dissertation had concentrated on the complex of changes involved in the emergence of analysis during the seventeenth century, and on the manifold ways in which the concept of proof evolved throughout that period; from early on he focused his attentions on one of the most enigmatic mathematicians of the period, Pierre de Fermat. When Mahoney's book on Fermat eventually took shape, it had the singular, and at the time unusual, feature in the history of mathematics of setting his subject squarely in the context of the period. Instead of mining Fermat's work for nuggets of future developments, Mahoney linked it with great care to the period's conceptions of the subject's central problems and techniques. Fermat, he wrote, "was a French mathematician of the first two-thirds of the seventeenth century. His thought, however original or novel, operated within a range of possibilities limited by that time and that place. His odyssey had its boundaries; his drummer beat to a tune of the times."[1]

As his research assistant at the time, Buchwald had the opportunity to learn directly from Mike as he gathered material on Fermat and worked hard to understand the logic behind the mathematics of, in his words, that "secretive and taciturn" man.[2] Many times one would enter Mike's study to find him surrounded by meticulously drawn diagrams and pages of elaborate geometry and analysis, all done in seventeenth-century fashion. He would occasionally break from his desk to enjoy flying a model plane in what were then the cornfields behind the university. On one occasion the plane disappeared into the field, never to be found, which reminded

Mike of the forever-lost pages on which Fermat had reportedly proved his famous theorem. His particular insights came strikingly together in a magisterial piece entitled "The Mathematical Realm of Nature," in which he allied mathematics with mechanics to show that "as a calculus of motion, analytic mechanics made motion a form of machine to be taken apart and reassembled. In that calculus, created at the turn of the eighteenth century, the new mechanics and the new mathematics met to form a new metaphysics."[3]

In the mid-1960s anyone wanting to use a computer at Princeton punched FORTRAN programs onto a series of cards and then, having fed the card stack into a reader, waited for the printout on the ubiquitous green-striped sheets of perforated paper. At the time Princeton's machine was an IBM 7094 which had by today's standards an utterly insignificant memory in the form of diminutive magnetic cores equivalent to about 144K. Mahoney was even then interested in mechanical computation, and he encouraged students to use the machine. In 1969 Princeton installed the first time-sharing arrangement, based on the IBM System 360, which allowed both input and output from remote IBM Selectric typewriters. Mahoney had by then recognized the revolutionary importance of the new procedures, and that summer Buchwald had a chance to work with him in exploring their potential.

Despite all this, Mike tended to be self-deprecatory about his own very early engagement with computer programming, and he enjoyed joking about how irritating the machines could be. At a 2004 lecture at the Center for Computing in the Humanities, at King's College, London, he told the crowd:

> During my final year at Harvard in 1959–60, I had a job as a computer programmer for a small electronics firm in Boston. It involved writing code for a Datatron 204, soon to become through acquisition the Burroughs 204, a decimally addressed, magnetic drum machine. Programming it meant understanding how it worked, since it was just you and the computer: no operating system, no programming support. Six or seven months of that persuaded me that computers were not very interesting, nor did they seem to me to have much of a future. So I abandoned my thoughts of going into applied mathematics and became a historian instead. With foresight like that, it was probably a good choice.[4]

But the truth was, despite his puckish anecdote, "mike@princeton.edu" (a handle that attested to his there-at-the-founding place in the digital university) never really stopped tinkering with computers and the code that makes them run: long before the World Wide Web existed he was conducting online discussion sessions for his classes by e-mail; he helped configure the first workstations for nonscientists at the residential colleges at Princeton; and by 1980 Mahoney's own research interests turned directly to the technoscientific revolution unfolding all around him—the dawning of the age of the "personal" computer. Mike was always justifiably proud of his ability to talk turkey with the current crop of computing techies on campus, and his enthusiasm for this sort of active engagement with technology was more than an avocation, it went to the heart of the questions that drove his intellectual life: head and hand, the abstract and the concrete, the mind and the machine. Threading between these hoary antitheses throughout his career, Mahoney sought again and again to understand the means by which history's finest thinkers left the stuff of the world behind, only in order to find it again, transformed.

His earliest work on the shift from geometrical to algebraic mathematics in the early modern period is perhaps the hallmark of this research, since the episode represents one of the mythical moments in the history of "abstraction." Moreover, Mahoney's later research on the origins of the programmable computer can be understood as something like the obverse of those very first problems to which he turned his attention: if algebraic mathematics had disembodied the world of inked lines and circles, the world of compass and straight edge, the history of twentieth-century computing amounted to something like the "reincarnation" of mathematical operations and their crystalline logic. A Turing machine is, in the end, not a *machine* at all, in much the same way that Descartes's hyperbolic lens-grinding machine in *La Dioptrique* was a fantastic illustration of the geometry of conic sections, but a mechanical fantasy. These slips—from logic, to heuristic, to model, to (if the engineers would cooperate) actual device—preoccupied Mahoney for close to half a century, and led him to deep interest in the history of technology and handcraft, an interest nevertheless inextricable from his grounding in the history of mathematics. Huygens's sea-clock, for instance, sat irresistibly at the nexus of these seemingly diverse concerns, since its gambit for resolving the consummately practical problem of the longitude hinged (literally) on a peculiar feature of the geometry of the cycloid (which is its own evolute), and required materializing the mathematical operation of the "evolution" of a curve by means of brass flanges and little bits of thread. It worked like a charm—but only, alas, on paper. Out of brilliant stories like these, brilliantly told, Mahoney explored the myriad ways that reason has fantasized its own transcendence, and craft has picked up the pieces. The result, as Mahoney taught in decades of well-attended lectures and intimate graduate seminars, is what we call modernity. To understand it, he argued, required rolling up your sleeves: many scholars have their students over to the house for tea or supper; only Mike had them over to use his workshop to reconstruct working models of several of the optical devices of Ibn-al-Hazen.

Both of us first encountered the field through Mike's remarkable lectures. As he once put it to a younger colleague: "the students have a right to see you think up there," and to that end he was sparing about his notes, and worked every year and every week to steep himself afresh in the primary sources that were the mainstay of his pedagogy. So exhilarating was it to watch him inhabit lost learning that one of us recalls feeling as a freshman, scribbling madly in the closing minutes of the final lecture of History 291, a vertiginous collapse of past, present, and future redolent of the sibylline climax of García Márquez's *Hundred Years of Solitude*: one felt, for a moment, perched on a fixed seat in Dickinson Hall, that the whole history of thought and action was collapsing on this very instant; that if we were patient for one more moment, the archive would become prophecy. It didn't, exactly; but then again, it did, since both of us had found our callings. And we were not alone.

Sadly now, we are.

NOTES / ACKNOWLEDGMENTS / INDEX

Notes

Unexpected Connections, Powerful Precedents, and Big Questions

1. In return he liked to call me a "sociologist," a reference to the somewhat misleading inclusion of this discipline in the title of the University of Pennsylvania's History and Sociology of Science department, in which I was a graduate student when we first met. Joseph November, one of his graduate assistants, also recalls using the phrase "old school" to describe Mahoney, this time in respect to Mahoney's "grades and expectations." "Usually," continued November, "that got rid of a few students" (J. November, personal communication, May 1, 2010).

2. Google Scholar currently identifies forty-nine citations to the paper. An examination of Google's reference counts for articles published in *Annals of the History of Computing* suggests that several articles by famous computing pioneers (for example, John Backus and Herman Goldstine) have been cited more frequently but that Mahoney's is the most widely cited article written by a professional historian.

3. His involvement was preceded by that of the veteran Harvard scholar I. Bernard Cohen, who did much to encourage early work in the area. Cohen's personal research on computing, however, was closely intertwined with his admiration for Harvard computing pioneer Howard Aiken and did not lead to any publicly articulated research agenda for the field.

4. Thomas S. Kuhn, *The Structure of Scientific Revolutions*, 2nd ed. (Chicago: University of Chicago Press, 1969). Kuhn was, as mentioned later in this introduction, Mahoney's dissertation adviser.

5. "Kuhn told us, don't listen to what they say. Watch what they do. Now, inventors and engineers don't talk much. Theirs is not a literate enterprise, so we're forced to watch what they do, at least insofar as you can reconstruct what they do from the artifacts." Michael S. Mahoney, *The Web of Learning:*

Staying Real in a Virtual World—Lecture Delivered at the University of South Carolina in Celebration of the College of Arts and Sciences, 6 April 2005 (accessed April 17, 2010), http://www.princeton.edu/~hos/Mahoney/articles/weboflearning/weboflearning.html. Mahoney was well familiar with the subsequent literature on the topic and in fall 2006 taught a graduate seminar called "Thinking Through Things and Thinking Things Through: The Material Epistemology of Science."

6. Mahoney apparently saw his early work on software as the basis for a book on the software crisis and software engineering. The online transcript of an interview with Berk Tague, as part of Mahoney's project on the history of UNIX, included Mahoney saying, "The major book I'm working on is the origin of the software crisis as it emerged in the '60s and how the industry got into that situation." http://www.princeton.edu/~hos/mike/transcripts/tague.htm. Unfortunately the transcript is not dated, but it appears to be part of a batch conducted in or just before 1989.

7. One approach might be to follow the agenda established by the 1980s within the science studies field to "open the black box" of technology and make connections between the apparently arcane inner workings of technological systems and the social contexts in which they were created. This is associated particularly with the work of Donald MacKenzie. MacKenzie's *Inventing Accuracy: A Historical Sociology of Nuclear Missile Guidance* (Cambridge, MA: MIT Press, 1990) was followed by his research on computing, published as *Mechanizing Proof* (Cambridge, MA: MIT Press, 2001) and *Knowing Machines* (Cambridge, MA: MIT Press, 1996). Within the history of technology a strong stream of work focused on practice can be traced to the concern within labor history for the organization of work practices on the shop floor. A good example, with a focus on computerization, is David F. Noble, *Forces of Production: A Social History of Industrial Automation* (New York: Alfred A. Knopf, 1984). Some attention to the relationship between systems software and applications programming practice is given in Thomas Haigh, "How Data Got Its Base: Generalized Information Storage Software in the 1950s and 60s," *IEEE Annals of the History of Computing* 31, no. 4 (October–December 2009): 6–25, and a detailed reconstruction of very early programming practice in Martin Campbell-Kelly, "The Development of Computer Programming in Britain (1945–1955)," *Annals of the History of Computing* 4, no. 2 (April 1982): 121–139, and related articles.

8. Michael S. Mahoney, "Reading a Machine," June 21, 2003, http://www.princeton.edu/~hos/h398/readmach/modeltfr.html (accessed March 18, 2010).

9. The discussion of the software crisis in the leading overview of the history of computing, Martin Campbell-Kelly and William Aspray, *Computer: A History of the Information Machine* (New York: Basic Books, 1996), 200–203, adopts a similar perspective and uses some of the same quotes. Likewise, Nathan Ensmenger and William Aspray, "Software as Labor Process," in *Mapping the History of Computing: Software Issues*, ed. Ulf Hashagen, Reinhard Keil-Slawik, and Arthur L. Norberg (New York: Springer-Verlag, 2002), 139–165, follows Mahoney's 1990 paper closely in taking the "soft-

ware crisis" as the central event in software history, positioning the 1968 NATO conference as a seminal event for programming practice, exploring the rhetoric of an industrial revolution in programming, and evaluating software engineering in the context of Taylorism. The NATO conference also looms large in another keynote contribution to the same volume: Donald MacKenzie, "A View from the Sonnenbichl: On the Historical Sociology of Software and System Dependability," in Hashagen, Keil-Slawik, and Norberg, *Mapping the History of Computing*, 97–122.

10. While MacKenzie's work on the software crisis was published after Mahoney's 1988 talk and subsequent 1990 paper, one of his students had already completed a dissertation with a chapter addressing the topic. Maria Eloina Pelaez Valdez, "A Gift from Pandora's Box: The Software Crisis" (Ph.D. diss., University of Edinburgh, 1988). Ideas similar to Mahoney's are also explored in Stuart S. Shapiro, "Computer Software as Technology: An Examination of Technological Development" (Ph.D. diss., Carnegie Mellon University, 1990).

11. Michael S. Mahoney, "What Makes the History of Software Hard," *IEEE Annals of the History of Computing* 30, no. 3 (July–September 2008): 8–18. This paper is not included in this collection because of its very substantial overlap with "Histories of Computing(s)."

12. Jean E. Sammett, "Introduction of Michael S. Mahoney," in *History of Programming Languages—II*, ed. Thomas J. Bergin and Richard G. Gibson (New York: ACM Press, 1996).

13. The National Science Foundation is explored in William Aspray and Bernard O. Williams, "Arming American Scientists: NSF and the Provision of Scientific Computing Facilities for Universities, 1950–73," *IEEE Annals of the History of Computing* 16, no. 4 (Winter 1994): 60–74. The work of DARPA/ARPA (Defense Advanced Research Projects Agency / Advanced Research Projects Agency) is documented in Arthur L. Norberg and Judy E. O'Neill, *Transforming Computer Technology: Information Processing for the Pentagon, 1962–1986* (Baltimore: Johns Hopkins University Press, 1996), and Alex Roland and Philip Shiman, *Strategic Computing: DARPA and the Quest for Machine Intelligence* (Cambridge, MA: MIT Press, 2002).

14. The most significant of which are Andrew Hodges, *Alan Turing: The Enigma of Intelligence* (New York: Simon and Schuster, 1983), and William Aspray, *John von Neumann and the Origins of Modern Computing* (Cambridge, MA: MIT Press, 1990).

15. That said, there have been two important investigations of the history of attempts to formally verify the correctness of programs against mathematical specifications, an effort rooted in the conceptual developments Mahoney described. One was by the sociologist of science Donald MacKenzie, whose book followed a similar approach to Mahoney in tying the emergence of this movement to theoretical developments of the 1960s and the 1968 NATO Conference on Software Engineering but focused primarily on attempts to put the approach into practice during the 1970s and 1980s: MacKenzie, *Mechanizing Proof*. The other is from Cliff Jones, himself an important figure in the story: "The Early Search for Tractable Ways of Reasoning about

Programs," *IEEE Annals of the History of Computing* 25, no. 2 (April 2003): 26–49.

16. For examples of these "impact trace graphs," see Wolfgang Emmerich, Mikio Aoyama, and Joe Sventek, "The Impact of Research on the Development of Middleware Technology," *ACM Transactions on Software Engineering and Methodology* 17, no. 4 (August 2008): 19-11–19-48.

17. Wolfgang Emmerich, e-mail message to author, May 9, 2010.

18. Michael S. Mahoney, "The Beginnings of Algebraic Thought in the Seventeenth Century," in *Descartes: Philosophy, Mathematics and Physics*, ed. S. Gaukroger (Sussex: Harvester Press, 1980), chap. 5.

19. Michael S. Mahoney, "Calculation—Thinking—Computational Thinking: Seventeenth-Century Perspectives on Computational Science," in *Form, Zahl, Ordnung: Studien zur Wissenschafts- und Technikgeschichte. Ivo Schneider zum 65. Geburtstag*, ed. Menso Folkerts and Rudolf Seising (Stuttgart: Frank Steiner Verlag, 2003).

20. Thomas S. Kuhn, "Second Thoughts on Paradigms," in *The Essential Tension: Selected Studies in Scientific Tradition and Change* (Chicago: University of Chicago Press, 1979), 293–319.

1. The History of Computing in the History of Technology

1. Fritz Machlup and Una Mansfeld, *The Study of Information: Interdisciplinary Messages* (New York: Wiley, 1983).

2. To characterize the unprecedented capabilities of computers linked to telecommunications, Nora and Minc coined the term *télématique* in *L'Informatisation de la Société* (Paris: La Documentation Française, 1978); translated into English as *The Computerization of Society* (Cambridge, MA: MIT Press, 1980).

3. For a recent, brief survey of the state of the field, see W. Aspray, "Literature and Institutions in the History of Computing," *ISIS* 75 (1984): 162–170. Many of the articles in *Computing Surveys*, begun in 1969, include a historical review of computing's history. Additionally see Saul Rosen, *Programming Systems and Languages* (New York: McGraw-Hill, 1967); Jean Sammet, *Programming Languages: History and Fundamentals* (Englewood Cliffs, NJ: Prentice Hall, 1969); compilations of papers such as B. Randell, ed., *Origins of Digital Computers: Selected Papers*, 3rd ed. (Berlin: Springer-Verlag, 1982); E. Yourdon, *Classics of Software Engineering* (New York: Yourdon Press, 1979); E. Yourdon, *Papers of the Revolution* (New York: Yourdon Press, 1982); and AT&T Bell Laboratories, *UNIX System Readings and Applications*, 2 vols. (Englewood Cliffs, NJ: Prentice Hall, 1987). Additionally, the twenty-fifth-anniversary issues of the leading journals also contain useful collections of important articles. Some examples of retrospectives include R. L. Wexelblatt's record of the 1978 ACM Conference on the History of Programming Languages, *History of Programming Languages* (New York: Academic Press, 1981), and a recent issue of the *Annals of the History of Computing* on the Burroughs B5000. C. J. Bashe, L. R. Johnson, J. H. Palmer, and E. W. Pugh, *IBM's Early Computers* (Cambridge, MA: MIT Press, 1986).

4. N. Stern, *From ENIAC to UNIVAC: An Appraisal of the Eckert-Mauchly Computers* (Billerica, MA: Digital Press, 1981); P. E. Ceruzzi, *Reckoners: The Prehistory of the Digital Computer, From Relays to the Stored Program Concept, 1935–1945* (Westport, CT: Greenwood Press, 1982); and M. Williams, *A History of Computing Technology* (Englewood Cliffs, NJ: Prentice Hall, 1986).

5. George Daniels put the question as an assertion (p. 6): "The real effect of technical innovation [has been] to help Americans do better what they had already shown a marked inclination to do." The seeming "social lag" in adapting to new technology, he argued, is more likely economic in nature. G. Daniels, "The Big Questions in the History of American Technology," *Technology and Culture* 11 (1970): 1–21.

6. N. Rosenberg, "Technological Interdependence in the American Economy," *Technology and Culture* 20 (1979): 25–50; and T. P. Hughes, *Networks of Power: Electrification in Western Society, 1880–1930* (Baltimore: Johns Hopkins University Press, 1983).

7. M. R. Smith, *Harper's Ferry Armory and the New Technology* (Ithaca, NY: Cornell University Press, 1977), and D. A. Hounshell, *From American System to Mass Production: The Development of Manufacturing Technology in the United States* (Baltimore: Johns Hopkins University Press, 1984). D. Nelson, *Managers and Workers: Origins of the New Factory System in the United States, 1880–1920* (Madison: University of Wisconsin Press, 1975), and S. Meyer, *The Five-Dollar Day* (Albany, NY: State University Press, 1981).

8. W. A. McDougall, *The Heavens and the Earth: A Political History of the Space Age* (New York: Basic Books, 1985). D. Noble, *Forces of Production: A Social History of Industrial Automation* (New York: Alfred A. Knopf, 1984). R. Cowan, *More Work for Mother: The Ironies of Household Technology from the Open Hearth to the Microwave* (New York: Basic Books, 1983).

9. A. F. P. Wallace, *Rockdale: The Growth of an American Village in the Early Industrial Revolution* (New York: Alfred A. Knopf, 1978), and E. Ferguson, "The Mind's Eye: Nonverbal Thought in Technology," *Science* 197 (1979): 827–836. See in particular Wallace's "Thinking about Machinery" (Wallace, *Rockdale*, pp. 237-240). B. Hindle, *Emulation and Invention* (New York: Basic Books, 1981); R. Jenkins, "Words, Images, Artifacts and Sound: Documents for the History of Technology," *British Journal for the History of Science* 20 (1987): 39–56. The quotation is from H. Ford, *My Life and Times* (Garden City, NY: Doubleday, 1922), 24. In *The Sciences of the Artificial*, Herbert Simon argues forcefully for the empirical nature of computer research that underlies its mathematical trappings. The thinking of computer designers and programmers is embodied in the way their machines and programs *work*, and the languages they use to specify how things are to work are themselves artifacts. The models they use are filled with images difficult or distractingly tedious to translate into words; H. Simon, *The Sciences of the Artificial*, 2nd ed. (Cambridge, MA: MIT Press, 1981). See also A. Newell and H. Simon, "Computer Science as Empirical Inquiry," *Communications of the ACM* 19 (1976): 113–126; and J. D. Bolter, *Turing's Man* (Chapel Hill: University of North Carolina Press, 1984).

10. I do not make this claim in ignorance of Konrad Zuse's Z4 or Alan Turing's ACE, which realized roughly the same goals as von Neumann's along independent paths. Clearly the computer was "in the air" by the 1940s. But it was the 1940s, not the 1840s.

11. I am including the history of mathematical logic in the history of mathematics.

12. It should sharpen the question for the history of science as well, if only by giving special force to the reciprocal influence of scientific theory and scientific instrumentation. But up to now at least it has not attracted the same attention. The computer may well change that as the shaping of scientific concepts and the pursuit of scientific inquiry come to depend on the state of computer technology.

13. P. Naur and B. Randell, eds., *Software Engineering: Report on a Conference Sponsored by the NATO Science Committee, Garmisch, Germany, 7th to 11th October, 1968* (Brussels: Scientific Affairs Divison, NATO, 1969).

14. Rosenberg, "Technological Interdependence," 26; Daniels, "Big Questions," 11.

15. J. Backus, "Can Programming Be Liberated from the Von Neumann Style? A Functional Style and Its Algebra of Programs," *Communications of the ACM* 21, no. 8 (1977): 613–641.

16. Elting E. Morison pursued this point along slightly different but equally revealing lines; E. E. Morison, *From Know-How to Nowhere: The Development of American Technology* (New York: Basic Books, 1974); Daniels, "Big Questions."

17. Lundstrom has recently chronicled the failure of some companies to make the requisite adjustments; D. E. Lundstrom, *A Few Good Men from Univac* (Cambridge, MA: MIT Press, 1987).

18. T. Kidder, *The Soul of a New Machine* (New York: Little, Brown and Co., 1981). The obvious citations here are Kraft and Greenbaum, but both works are concerned more with politics than with computing, and the focus of their political concerns, the "deskilling" of programmers through the imposition of methods of structured programming, has proved ephemeral, as subsequent experience and data show that programmers have made the transition with no significant loss of control over their work. P. Kraft, *Programmers and Managers: The Routinization of Computer Programming in the United States* (New York: Springer-Verlag, 1977); J. Greenbaum, *In the Name of Efficiency: Management Theory and Shopfloor Practice in Data Processing Work* (Philadelphia: Temple University Press, 1979); see also B. Boehm, *Software Engineering Economics* (Englewood Cliffs, NJ: Prentice Hall, 1981).

19. J. H. Wilkinson, "Some Comments from a Numerical Analyst," *Journal of the ACM* 18, no. 2 (1971): 137–147. See, for example, Burke: "Thus technological innovation is not the product of society as a whole but emanates rather from certain segments within or outside of it; the men or institutions responsible for the innovation, to be successful, must 'sell' it to the general public; and innovation *does* have the effect of creating broad social change";

J. G. Burke, "Comment: The Complex Nature of Explanation in the Historiography of Technology," *Technology and Culture* 11 (1970): 23. Ferguson has made a similar observation about selling new technology; T. S. Ferguson, "The American-ness of American Technology," *Technology and Culture* 20 (1979): 3–24. Kidder, *Soul of a New Machine*.

20. C. C. Herrmann and J. F. Magee, "'Operations Research' for Management," *Harvard Business Review* (July–August 1953): 100–112. J. Diebold, "Automation—the New Technology," *Harvard Business Review* (November–December 1953): 63–71. H. A. Simon, *The New Science of Management Decision* (New York: Harper and Row, 1960), 14.

21. M. L. Hurni, "Decision Making in the Age of Automation," *Harvard Business Review* 34 (September–October 1955): 49.

22. Along these lines, historians of computing would do well to remember that a line of writings on the nature, impact, and even history of computing stretching from Edmund C. Berkeley's *Giant Brains* through John Diebold's several volumes to Edward Feigenbaum's and Pamela McCorduck's *The Fifth Generation* stems from people with a product to sell, whether management consulting or expert systems. E. C. Berkeley, *Giant Brains or Machines That Think* (New York: Wiley, 1949), and E. Feigenbaum and P. McCorduck, *The Fifth Generation: Artificial Intelligence and Japan's Computer Challenge to the World* (Reading, MA: Addison-Wesley, 1983).

23. F. Rose, *Into the Heart of the Mind: An American Quest for Artificial Intelligence* (New York: Random House, 1984), 36.

24. H. Rheingold, *Tools for Thought: The People and Ideas Behind the Next Computer Revolution* (New York: Simon and Schuster, 1985).

25. An effort at international cooperation in establishing a standard programming language, ALGOL from its inception in 1956 to its final (and, some argued, over-refined) form in 1968 provides a multileveled view of computing in the 1960s. While contributing richly to the conceptual development of programming languages, it also has a political history which carries down to the present in differing directions of research, both in computer science and, perhaps most clearly, in software engineering.

26. Kidder, *Soul of a New Machine*.

27. G. M. Weinberg, *The Psychology of Computer Programming* (New York: Van Nostrand Reinhold, 1971), Chapter 1. R. DeMillo, R. J. Lipton, and A. J. Perlis, "Social Processes and Proofs of Theorems and Programs," *Communications of the ACM* 22, no. 5 (1979): 271–280.

28. Daniels, "Big Question."

29. M. D. McIlroy, "Mass-Produced Software Components," in Naur and Randell, *Software Engineering*.

30. One has to wonder about an article on software engineering that envisions progress on an industrial model and uses photographs taken from the Great Depression.

31. N. Rosenberg, "Technological Change in the Machine Tool Industry, 1840–1910," *Journal of Economic History* 23 (1963): 414–443. Hounshell, *American System to Mass Production*.

32. The latter designation stems from E. Frand, "Thoughts on Product Development: Remembrance of Things Past," *Industrial Research and Development* (July 1983): 23.
33. R. S. Lynd and H. M. Lynd, *Middletown* (New York: Harcourt, Brace and World, 1929).
34. I. B. Cohen, *Revolutions in Science* (Cambridge, MA: Harvard University Press, 1986).

2. What Makes History?

1. Richard Hamming, "We Would Know What They Thought When They Did It," in *A History of Computing in the Twentieth Century: A Collection of Essays*, ed. N. Metropolis, J. Howlett, and G.-C. Rota, 3–9 (New York: Academic Press, 1980).

3. Issues in the History of Computing

1. Grace Hopper, "Business Data Processing—A Review," *IFIP* 62 (1962): 35–39.
2. Thomas S. Kuhn, *The Structure of Scientific Revolutions* (Chicago: University of Chicago Press, 1962).
3. Tracy Kidder, *The Soul of a New Machine* (Boston: Little, Brown, and Co., 1981), 26.
4. Eugene S. Ferguson, "The Mind's Eye: Nonverbal Thought in Technology," *Science* 197 (1979): 827–836; Anthony F. C. Wallace, "Thinking About Machinery," in *Rockdale: The Growth of an American Village in the Early Industrial Revolution* (New York: Alfred A. Knopf, 1978), 237. The quotation is from Henry Ford, *My Life and Work* (Garden City, NY: Doubleday, 1922), 23–24.
5. Richard W. Hamming, "We Would Know What They Thought When They Did It," in *A History of Computing in the Twentieth Century: A Collection of Essays*, ed. N. Metropolis, J. Howlett, and G.-C. Rota, 3–9 (New York: Academic Press, 1980).
6. Alan C. Kay, "The Early History of Smalltalk," *SIGPLAN Notices* 28, no. 3 (March 1993): 71.
7. Gerald M. Weinberg, *The Psychology of Computer Programming* (New York: Van Nostrand Reinhold Co., 1979); Brian Kernighan and P. J. Plauger, *The Elements of Programming Style* (New York: McGraw-Hill, 1974).
8. Derek J. de Solla Price, *Little Science, Big Science* (New York: Columbia University Press, 1963).
9. Alan Perlis, "The American Side of the Development of Algol," in *History of Programming Languages*, ed. Richard L. Wexelblat, (New York: Academic Press, 1981), 83.
10. Perlis, "Algol," 164–165.
11. Kenneth Flamm, *Creating the Computer: Government, Industry, and High Technology* (Washington, DC: Brookings Institution, 1988). Arthur L. Norberg

and Judy E. O'Neill, *Promoting Technological Innovation: The Information Processing Techniques Office of the Defense Advanced Research Projects Agency*, Report to the Software and Intelligent Systems Technology Office, DARPA (Minneapolis, MN: Charles Babbage Institute, 1992). Charles J. Bashe, Lyle R. Johnson, John H. Palmer, and Emerson W. Pugh, *IBM's Early Computers* (Cambridge, MA: MIT Press, 1986).

12. Cyril C. Herrmann and John F. Magee, "'Operations Research' for Management," *Harvard Business Review* 31, no. 4 (1953): 100–112; first advertisements for IBM and Univac in *Harvard Business Review* 31, 5–6 (1953).

13. Bjarne Stroustrup, "A History of C++: 1979–1991," *SIGPLAN Notices* 28, no. 3 (March 1993): 294.

14. Grace Hopper, "The Education of a Computer," *Proceedings of the Association for Computing Machinery Conference, Pittsburgh, May 1952*; repr. with introduction by David Gries, *Annals of the History of Computing* 9, nos. 3–4 (1988): 271–281; Peter Wegner, "Capital-Intensive Software Technology," *IEEE Software* 1, no. 3 (1984): 7–45; Mary Shaw, "Prospects for an Engineering Discipline of Software," *IEEE Software* 7, no. 6 (November 1990): 15–24.

15. M. D. McIlroy, "On Mass-Produced Software Components," in *Software Engineering: Concepts and Techniques. Proceedings of the NATO Conferences*, ed. Peter Naur, Brian Randell, and J. N. Buxton, 88–95 (Garmisch, 1968; Rome, 1969; New York: Petrocelli/Charter, 1976).

16. Andrew Abbott, *The System of Professions: An Essay on the Division of Expert Labor* (Chicago: University of Chicago Press, 1988), 2–3.

17. Barry W. Boehm, "Software and Its Impact: A Quantitative Assessment," *Datamation* 19 (May 1973): 48–59.

18. Edmund C. Berkeley, *The Computer Revolution* (Garden City, NY: Doubleday and Co., 1962). Michael S. Mahoney, "The History of Computing in the History of Technology," *Annals of the History of Computing* 10 (1988): 113–125; Robert S. and Helen Lynd, *Middletown: A Study in American Culture* (New York: Harcourt, Brace and World, 1929; reprinted 1956).

4. The Histories of Computing(s)

1. Marx made the point in *Capital*, chap. 15 ("Machinery and Large-Scale Industry"), fn. 18: "To what extent the old forms of the instruments of production influence their new forms at the beginning is shown among other things, by the most superficial comparison of the present power-loom with the old one, of the modern blowing apparatus of a blast-furnace with the first inefficient mechanical reproduction of the ordinary bellows, and perhaps more strikingly than in any other way by the fact that, before the invention of the present locomotive, an attempt was made to construct a locomotive with two feet, which it raised from the ground alternately, like a horse. It is only after a considerable development of the science of mechanics, and an accumulation of practical experience, that the form of a machine becomes settled entirely in accordance with mechanical principles, and emancipated from the

traditional form of the tool from which it has emerged." Karl Marx, *Capital*, trans. Ben Fowkes (New York: Vintage Books, 1977), vol. 1, 505.

2. Edmund C. Berkeley, *The Computer Revolution* (Garden City, NY: Doubleday, 1962).

3. Thus the opening words of Steven Shapin's *The Scientific Revolution* (Chicago: University of Chicago Press, 1996): "There was no such thing as the Scientific Revolution, and this is a book about it" (1).

4. See, for example, Michael S. Mahoney, "Computer Science: The Search for a Mathematical Theory," in *Science in the 20th Century*, ed. John Krige and Dominique Pestre (Amsterdam: Harwood Academic Publishers, 1997), chap. 31. Michael S. Mahoney, "Finding a History for Software Engineering," *IEEE Annals of the History of Computing* 26, no. 1 (2004): 8–19.

5. Quoted in Thomas Haigh, "The Chromium-Plated Tabulator: Institutionalizing an Electronic Revolution, 1954–1958," *IEEE Annals of the History of Computing* 23, no. 4 (2001): 77. Despite the title, Haigh emphasizes the slow, evolutionary process of the transition from punched-card machinery to computer systems.

6. For that history, see inter alia Martin Davis, *The Universal Computer: The Road from Leibniz to Turing* (New York: W.W. Norton, 2000), and Sybille Krämer, *Symbolische Maschinen: Die Idee der Formalisierung in geschichtlichem Abriss* (Darmstadt: Wissenschaftliche Buchgesellschaft, 1988).

7. See, for example, Jennifer Tann, *The Development of the Factory* (London: Cornmarket Press, 1970), which is based on the papers of the Boulton and Watt Company.

8. SAGE—Semi-Automatic Ground Environment; WWMCCS—World-Wide Military Command and Control System; David A. Mindell, *Between Human and Machine: Feedback, Control, and Computing before Cybernetics* (Baltimore, MD: Johns Hopkins University Press, 2002).

9. James A. Cortada, *Information Technology as Business History: Issues in the History and Management of Computers* (Westport, CT: Greenwood Press, 1996), ix.

10. Haigh, "Chromium-Plated Tabulator," 75.

11. Ibid., 94.

12. Jon Agar, *The Government Machine: A Revolutionary History of the Computer* (Cambridge, MA: MIT Press, 2004), 3.

13. Ulf Hashagen, Reinhard Keil-Slawik, and Arthur Norberg, *History of Computing: Software Issues* (Berlin: Springer-Verlag, 2002).

14. James E. Tomayko, "Software as Engineering," in Hashagen, Keil-Slawik, and Norberg, *History of Computing*. Henry Petroski, *To Engineer Is Human: The Role of Failure in Successful Design* (New York: St. Martin's Press, 1985).

15. Peter Naur and Brian Randell, eds., *Software Engineering: Report on a Conference Sponsored by the NATO Science Committee, Garmisch, Germany, 7th to 11th October 1968* (Brussels: Scientific Affairs Division, NATO, 1969), 13. The report was republished, together with the report on the second conference in Rome the following year, in *Software Engineering: Concepts and Techniques. Proceedings of the NATO Conferences*, ed. Peter

Naur, Brian Randell, and J. N. Buxton (New York: Petrocelli, 1976). Randell has made both reports available for download in pdf format at http://homep ages.cs.ncl.ac.uk/brian.randell/NATO/.

16. Frederick P. Brooks, "No Silver Bullet—Essence and Accidents of Software Engineering," *Information Processing 86*, ed. H. J. Kugler (Amsterdam: Elsevier Science, 1986), 1069–1076; repr. in *Computer* 20, no. 4 (1987): 10–19; and in the Anniversary Edition of *The Mythical Man-Month: Essays on Software Engineering* (Reading, MA: Addison-Wesley, 1995), chap. 16. Chap. 17, "'No Silver Bullet' Refired" is a response to critics of the original article and a review of the silver bullets that have missed the mark over the intervening decade.

17. "Science of Design Funding," http://www.nsf.gov/funding/pgm_summ.jsp ?pims_id=12766.

18. Henry Ford, *My Life and Work* (Garden City, NY: Doubleday, 1922), 23–24.

19. Langdon Winner, *Autonomous Technology: Technics-out-of-Control as a Theme in Political Thought* (Cambridge, MA: MIT Press, 1977), 202: "We do not *use* technologies so much as *live* them." In a talk some years ago, Ruth Schwarz Cowan, author of *More Work for Mother*, related how living for a week in Bethpage Village, a historical reconstruction of an eighteenth-century homestead on Long Island, reshaped her relations with her daughters through the cooperative patterns of work required by the household technology of the time.

20. This situation has implications reaching beyond the history of computing. U.S. government records of the 1950s now sit on magnetic tapes that cannot be read because neither the software nor the hardware with which they were created has survived.

21. Christopher G. Langton, "Artificial Life" [1989], in *Philosophy of Artificial Life*, ed. Margaret Boden, 47 (Oxford: Oxford University Press, 1996).

22. Willard McCarty, "As It Almost Was: Historiography of Recent Things," *Literary and Linguistic Computing* 19, no. 2 (2004): 161–180. Jay David Bolter and Richard Grusin, *Remediation: Understanding New Media* (Cambridge, MA: MIT Press, 1999).

5. Software: The Self-Programming Machine

1. Werner Buchholz, "The System Design of the IBM 701 Computer," *Proceedings of the IRE* 41, no. 10 (1953): 1262–1275.

2. Herbert A. Simon, "The Corporation: Will It Be Managed by Machines?," published in 1961 in a volume of essays on *Management and Corporations: 1985*, ed. Melvin Anshen and George Leland Bach (New York: McGraw-Hill, 1961); repr. in *The World of the Computer*, ed. John Diebold (New York: Random House, 1973), 154.

3. Maurice V. Wilkes, *Memoirs of a Computer Pioneer* (Cambridge, MA: MIT Press, 1985), 145.

4. H. Rutishauser wrote in 1967 ("Description of Algol 60," *Handbook for Automatic Computation* [Berlin: Springer]) that "by 1954 the idea of using a computer for assisting the programmer had been seriously considered in Europe,

but apparently none of these early algorithmic languages was ever put to actual use. The situation was quite different in the USA, where an assembly language epoch preceded the introduction of algorithmic languages. To some extent this may have diverted attention and energy from the latter, but on the other hand it helped to make automatic programming popular in the USA" (quoted by Peter Naur in "The European Side of the Last Phase of the Development of Algol 60," *History of Programming Languages*, ed. Richard L. Wexelblat, 93 (New York: Academic Press, 1981).

5. Naur, "European Side," 95–96. In designing Algol 60, the members of the committee expressly barred discussions of implementation of the features of the language, albeit on the shared assumption that no one would propose a feature he did not know how to implement, at least in principle.

6. Peter Naur and Brian Randell, eds., *Software Engineering: Report on a Conference Sponsored by the NATO Science Committee, Garmisch, Germany, 7th to 11th October 1968* (Brussels: NATO Scientific Affairs Division, January 1969), 13.

7. R. W. Bemer, "Position Paper for Panel Discussion [on] the Economics of Program Production," *Information Processing 68* (Amsterdam: North-Holland Publishing Company, 1969), II, 1626.

6. Extracts from *The Roots of Software Engineering*

1. Published in N. Metropolis, J. Howlett, G.-C. Rota, eds., *A History of Computing in the Twentieth Century: A Collection of Essays* (New York: Academic Press, 1980), 3–9.

2. Brian Randell ("Software Engineering in 1968," *Proceeding of the 4th International Conference on Software Engineering* [Munich, 1979], 1) ascribes it to J. P. Eckert at the Fall Joint Computer Conference in 1965, but the transcript of the one panel discussion in which Eckert participated shows no evidence of the term "software engineering." D. T. Ross claims the term was used in courses he was teaching at MIT in the late '50s; cf. "Interview: Douglas Ross Talks About Structured Analysis," *Computer* (July 1985): 80–88. Peter Naur, Brian Randell, and J. N. Buxton, eds., *Software Engineering: Concepts and Techniques* (New York: Petrocelli/Charter, 1976).

3. See, for example, Michael R. Williams, *A History of Computing Technology* (Englewood Cliffs, NJ: Prentice Hall, 1986). M. S. Mahoney, "Computers and Mathematics: The Search for a Discipline of Computer Science," to appear in the *Proceedings of the International Symposium on Structures in Mathematical Theories*, San Sebastian-Donostia, Spain, September 1990. In a real sense, numerical analysis came into being with the computer. The term itself is of postwar coinage. B. V. Bowden, ed., *Faster Than Thought: A Symposium on Digital Computing Machines* (New York, 1953), 96–97.

4. See in particular Charles J. Bashe, Lyle R Johnson, John H Palmer, and Emerson W. Pugh, *IBM's Early Computers* (Cambridge, MA: MIT Press, 1986), and Kenneth Flamm, *Creating the Computer* (Washington, DC: Brookings Institution, 1988), for American developments and John Hendry, *Innovating*

for Failure: Government Policy and the Early British Computer Industry (Cambridge, MA: MIT Press, 1989), for contrasting efforts in Britain.

7. Finding a History for Software Engineering

1. Indeed, this article stems from just such an address, delivered to ACM SIG-SOFT's Ninth Foundations of Software Engineering Conference (FSEC 9) in 1998. The first quotation is from W. Humphrey, "The Software Engineering Process: Definition and Scope," *Representing and Enacting the Software Process: Proceedings of the 4th International Software Process Workshop* (New York: ACM Press, 1989), 82. The second is found in M. Shaw, "Prospects for an Engineering Discipline of Software," *IEEE Software* 7, no. 6 (November 1990): 15.
2. See, for example, A. D. Abbott, *The System of Professions: An Essay on the Division of Expert Labor* (Chicago: University of Chicago Press, 1988).
3. For a recent discussion of the question, see *The History of Software Engineering*, Report of the Dagstuhl Seminar No. 9635, ed. W. Aspray, R. Keil-Slawik, and D. Parnas (Dagstuhl, 1996), available online at http://www.dagstuhl.de/files/Reports/96/9635.pdf, and J. E. Tomayko, "Software as Engineering" with commentaries by A. Endres and B. E. Seely, *History of Computing: Software Issues*, ed. U. Hashagen, R. Keil-Slawik, and A. Norberg (Berlin: Springer-Verlag, 2002). Mary Shaw of Carnegie Mellon University and the Software Engineering Institute took this approach explicitly in "Prospects for an Engineering Discipline of Software," *IEEE Software* 7, no. 6 (November 1990): 15–24, where she proposed a historical model of the professionalization of engineering based primarily on the development of chemical engineering. Her diagram of the process reappeared in enhanced form in W. W. Gibbs, "Software's Chronic Crisis," *Scientific American* 271, no. 3 (September 1994): 86–95, at 92. For example, at the first NATO conference (see below), Ronald Graham of Bell Labs remarked that "we build systems like the Wright brothers built airplanes—build the whole thing, push it off the cliff, let it crash, and start over again" (*Software Engineering: Concepts and Techniques. Proceedings of the NATO Conferences*, ed. P. Naur, B. Randell, and J. N. Buxton [New York: Petrocelli, 1976], 7). Historians of technology know that the Wright Brothers' successful flight was in fact the culmination of a carefully planned, theoretically and empirically informed program of research and development. In particular, they had a relatively clear idea of what problems they had to solve and of how they might go about solving them. Whether or not their approach might have served as a useful example for fledgling software engineers, it does not seem prima facie to constitute a negative example.
4. P. Naur and B. Randell, eds., *Software Engineering: Report on a Conference Sponsored by the NATO Science Committee, Garmisch, Germany, 7th to 11th October 1968* (Brussels: Scientific Affairs Division, NATO, 1969), 13. The report was republished, together with the report on the second conference in Rome the following year, in *Software Engineering: Concepts and Techniques. Proceedings of the NATO Conferences*, ed. P. Naur, B. Randell,

and J. N. Buxton (New York: Petrocelli, 1976). Randell has made both reports available for download in pdf format at http://homepages.cs.ncl.ac.uk/brian.randell/NATO/. The site includes photographs of the participants and some of the sessions at Garmisch.

5. On the formation of the agendas of theoretical computer science, see M. S. Mahoney, "Software as Science—Science as Software," in *History of Computing: Software Issues*, ed. U. Hashagen, R. Keil-Slawik, and A. Norberg, 25–48 (Berlin: Springer-Verlag, 2002).

6. "Myth" here should be taken in the sense of a story told by a community to account for why it does things they way it does. The story may be more or less factually accurate, but its function does not depend on it.

7. Edsgar W. Dijksta was the foremost proponent of this "European" view. The quotation is from John McCarthy, "Towards a Mathematical Science of Computation," *Information Processing 1962: Proceedings of IFIP Congress 62, Munich, 1962* (IFIP 62) (Amsterdam: North-Holland, 1963), 21.

8. John McCarthy, "A Basis for a Mathematical Theory of Computation," *Proceedings of the Western Joint Computer Conf.* 19 (May 1961): 225–238; reprinted, with corrections and an added tenth section, in *Computer Programming and Formal Systems*, ed. P. Braffort and D. Hirschberg (Amsterdam: North-Holland, 1963), 33–70, at 69.

9. McCarthy, "Basis," 33.

10. Ibid., 34. McCarthy argued that none of the three current (1961) directions of research into the mathematics of computing held much promise of such a science. Numerical analysis was too narrowly focused. The theory of computability set a framework into which any mathematics of computation would have to fit, but it focused on what was unsolvable rather than seeking positive results, and its level of description was too general to capture actual algorithms. Finally, the theory of finite automata, though it operated at the right level of generality, exploded in complexity with the size of current computers. As he explained in another article, "The fact of finiteness is used to show that the automaton will eventually repeat a state. However, anyone who waits for an IBM 7090 to repeat a state, solely because it is a finite automaton, is in for a very long wait." ("Towards a Mathematical Science of Computation," *Proc. IFIP Congress 62* [Amsterdam: North-Holland, 1963], 22).

11. C. A. R. Hoare, "Programming: Sorcery or Science?," *IEEE Software* 1, no. 2 (March 1984): 5–16, at 10. Perhaps only coincidentally the article included a photograph of the room in which Kepler died (14).

12. For an overview, see M. S. Mahoney, "The Structures of Computation," in *The First Computers—Histories and Architectures*, ed. R. Rojas and U. Hashagen (Cambridge, MA: MIT Press, 2000).

13. Naur, Randell, and Buxton, *Software Engineering*, 147.

14. The committee's report was Bruce W. Arden, ed., *What Can Be Automated? (COSERS)* (Cambridge, MA: MIT Press, 1980), 139. The committee consisted of Richard M. Karp (chair; Berkeley), Zohar Manna (Stanford), Albert R. Meyer (MIT), John C. Reynolds (Syracuse), Robert W. Ritchie (Washington), Jeffrey D. Ullman (Stanford), and Shmuel Winograd (IBM Research).

B. Boehm, "Software Engineering," *IEEE Transactions on Computers* C-25, no. 12 (December 1976): 1226–1241, at 1226 (repr. in *Milestones of Software Engineering*, ed. P. W. Oman and Ted G. Lewis [Hoboken, NJ: IEEE Computer Society Press, 1990], 54–69, at 54). An early leader in the field of software metrics, Boehm later developed COCOMO, a system for estimating the cost of software projects, and wrote the leading text in the subject, *Software Engineering Economics*.

15. B. Boehm, "Software and Its Impact: A Quantitative Assessment," *Datamation* 19 (1973): 48–59.

16. Boehm, "Software Engineering," 67. Boehm's footnote to "technicians" is worth repeating here. "For example, a recent survey of 14 installations in one large organization produced the following profile of its 'average coder': 2 years college-level education, 2 years software experience, familiarity with 2 programming languages and 2 applications, and generally introverted, sloppy, inflexible, 'in over his head,' and undermanaged. Given the continuing increase in demand for software personnel, one should not assume that this typical profile will improve much. This has strong implications for effective software engineering technology which, like effective software, must be well-matched to the people who use it."

17. F. L. Bauer, "Software Engineering," *Information Processing* 71 (1972): 530–538, at 530. Repr. in *Advanced Course in Software Engineering*, ed. F. L. Bauer (Berlin: Springer-Verlag, 1973), 522–545; the reprint did not include Bauer's playful parody of a computer scientist's design of a three-prong hay fork.

18. M. D. McIlroy, "Mass Produced Software Components," in Naur and Randell, *Software Engineering*, 138–150, at 138–139. At the time, McIlroy was one of the representatives of Bell Labs to the Multics project at MIT, where he worked on the semantics of PL/I. He subsequently oversaw the development of UNIX, to which he contributed the notion of "pipes," which allows the chaining of programs, each taking as its input the output of its predecessor.

19. Nathan Rosenberg, "Technological Change in the Machine Tool Industry, 1840–1910," *Journal of Economic History* 23 (1963): 414–443; repr. in Rosenberg, *Perspectives on Technology* (Cambridge: Cambridge University Press, 1976), chap. 1. McIlroy in conversation at Bell Labs, fall 1989.

20. Jack B. Dennis, "Modularity," in Bauer, *Advanced Course in Software Engineering*, chap. 3.A, at 128.

21. Brad J. Cox, "Planning the Software Industrial Revolution," *IEEE Software* 7, no. 6 (November 1990). Peter Wegner, "Capital-Intensive Software Technology," *IEEE Software* 1, no. 3 (July 1984): 7–45. Both Wegner and Jones have told me that their editors, not they, chose the pictures in question. Thus, the analogy was widely shared in the larger community. Gregory W. Jones, *Software Engineering* (New York: Wiley, 1990).

22. F. L. Bauer, "Software Engineering," 532.

23. Ibid., 533.

24. R. W. Bemer, "Position Paper for Panel Discussion [on] the Economics of Program Production," *Information Processing 68* (Amsterdam: North-Holland Publishing Company, 1969), II, 1626.

25. In the now classic *Taylorism at Watertown Arsenal: Scientific Management in Action, 1908–1915* (Cambridge, MA: Harvard University Press, 1960; repr. as *Scientific Management in Action: Taylorism at Watertown Arsenal, 1908–1915* [Princeton, NJ: Princeton University Press, 1985]), H. G. J. Aitken listed Taylor's "solutions of enduring significance" (29): (1) the planned routing and scheduling of work in progress, leading to the assembly line and continuous flow production; (2) systematic inspection procedures between operations; (3) printed job and instruction cards; (4) refined cost-accounting techniques; (5) systematization of store procedures, purchasing, and inventory control; (6) and "functional foremanship," which was the only element not to gain general acceptance. Taylor got little credit from historians for these things, yet "these inconspicuous innovations have probably exercised a more far-reaching influence on industrial practice than has the conspicuous innovation of stop-watch time study." Taylor and Taylorism have attracted renewed attention from historians in recent decades; see in particular D. Nelson, ed., *A Mental Revolution: Scientific Management Since Taylor* (Columbus: Ohio State University Press, 1992), and S. P. Waring, *Taylorism Transformed: Scientific Management since 1945* (Chapel Hill: University of North Carolina Press, 1991). R. Kanigel's *The One Best Way: Frederick Winslow Taylor and the Enigma of Efficiency* (New York: Viking Press, 1997) is a full and informative biography.

26. That science constituted the famous "one best way" on which Taylor's system rested. F. W. Taylor, *The Principles of Scientific Management* (1911; repr., New York: Norton, 1967), 36–37.

27. W. W. Agresti, "Software Engineering as Industrial Engineering," *Software Engineering Notes* 6, no. 5 (1981): 11–12, at 11. I thank Michael Cusumano for drawing my attention to this article. Agresti later moved to Computer Sciences Corporation and then to MITRE Corporation.

28. L. J. Osterweil, "Software Processes Are Software Too," *Proceedings of the Ninth International Conference on Software Engineering* (ICSE 9) (Hoboken, NJ: IEEE Computer Society Press, 1987), 2–13. At ICSE 19, Osterweil's paper was recognized as the most influential paper of ICSE 9. W. S. Humphrey, "The Personal Software Process: Status and Trends," *IEEE Software* 17, no. 6 (November–December 2000): 72.

29. F. P. Brooks, Jr., *The Mythical Man-Month: Essays on Software Engineering* (Reading, MA: Addison-Wesley, 1975), 47–48.

30. D. H. Brandon, "The Economics of Computer Programming," in *On the Management of Computer Programming*, ed. G. F. Weinwurm (New York: Auerbach, 1970), chap. 1. Brandon evidently viewed management through Taylorist eyes, but he was clear-sighted enough to see that computer programming failed to meet the prerequisites for scientific management. For an analysis of why testing was so unreliable, see R. N. Reinstedt, "Results of a Programmer Performance Prediction Study," *IEEE Transactions on Engineering Management* (December 1967): 183–187, and G. M. Weinberg, *The Psychology of Computer Programming* (New York: Dorset House, 1971), chap. 9.

31. M. Cusumano, *Japan's Software Factories*, (New York: Oxford University Press, 1991), 147–148; the quotation is from the description of the SDC software factory by H. Bratman and T. Court, "Elements of the Software Factory: Standards, Procedures, and Tools," in *Software Engineering Techniques* (Maidenhead: Infotech International, 1977), 137. For a historical overview of the concept, see Cusumano, "Shifting Economies: From Craft Production to Flexible Systems and Software Factories," *Research Policy* 21 (1992): 453–480.

32. H. Ford, *My Life and Work* (New York: Doubleday, 1922), 82–83.

33. See B. Blum, "Understanding the Software Paradox," *ACM SIGSOFT Software Engineering Notes* 10, no. 1 (1985): 43–47, who attributes the notion to L. G. Stucki.

34. J. N. Buxton and B. Randell, eds., *Software Engineering Techniques: Report on a Conference Sponsored by the NATO Science Committee, Rome, Italy, 27th to 31st October 1969* (Brussels: NATO Science Committee, 1969), 12. For a review of the architectural model, see J. O. Coplien, "Reevaluating the Architectural Metaphor: Toward Piecemeal Growth," *IEEE Software* 16, no. 5 (September–October 1999): 40–44.

8. Boys' Toys and Women's Work: Feminism Engages Software

1. Pamela E. Mack, "What Difference Has Feminism Made to Engineering in the Twentieth Century" in *Feminism in Twentieth-Century Science, Technology, and Medicine*, ed. Angela N. H. Creager, Elizabeth Lunbeck, and Londa Schiebinger (Chicago: University of Chicago Press, 2001), 149–168.

2. Her name lives on in particular in the Grace Murray Hopper Award of the Association of Computing Machinery (for outstanding achievement by a researcher under thirty; established 1971, not yet awarded to a woman) and in the destroyer USS *Hopper*, commissioned in 1997.

3. W. Barkley Fritz, "The Women of ENIAC," *IEEE Annals of the History of Computing* 18, no. 3 (1996): 13–28, and Jennifer Light, "When Computers Were Women," *Technology and Culture* 4, no. 3 (1999): 455–483. On the loss of technical jobs following the war, see Margaret Rossiter, *Women Scientists in America: Before Affirmative Action, 1940–1972* (Baltimore: Johns Hopkins University Press, 1995). The Electronic Numerical Integrator And Calculator went into operation on 14 February 1946 at the University of Pennsylvania's Moore School of Engineering. It was the basis for the subsequent development of the stored-program device now commonly understood as the electronic computer. The quotation is from John Backus, "Programming in America in the 1950s—Some Personal Impressions," in *A History of Computing in the Twentieth Century*, ed. N. Metropolis et al. (New York: Academic Press, 1980), 125–135.

4. On the emergence of software, see Michael S. Mahoney, "Software: The Self-Programming Machine," in *From 0 to 1: An Authoritative History of Modern Computing*, ed. Atsushi Akera and Frederik Nebeker (Oxford: Oxford University Press, 2002), 91–100.

5. For the first round of studies, see *Sex Roles* 13, nos. 3–4 (1985), a special issue on "Women, Girls, and Computers," which includes empirical studies ranging from kindergarten to adults in the office. Subsequent studies have in general reinforced the basic tendency, but with some conflicting findings. For a critical evaluation of seven such reports, see Robin Kay, "An Analysis of Methods Used to Examine Gender Differences in Computer-Related Behavior," *Journal of Educational Computing Research* 8, no. 3 (1992): 277–279; and for a recent effort to disaggregate the factors involved, see Lori J. Nelson and Joel Cooper, "Gender Differences in Children's Reactions to Success and Failure with Computers," *Computers in Human Behavior* 13, no. 2 (1997): 247–267. Despite the conflicts and critiques, the general trend of these results would seem to have profound implications for educational policy, but few commentators, much less school boards, seem to have considered the effects of introducing into classrooms devices that students so clearly associated with one gender rather than the other. Interestingly, the question does not appear to have caught the attention of the current administration in Washington, perhaps because it places its concern for women's issues in conflict with its enthusiasm for educational technology. [Added 2004] For an extended discussion of the issues and literature, see now Jane Margolis and Allan Fisher, *Unlocking the Clubhouse: Women in Computing* (Cambridge, MA: MIT Press, 2002).

6. Perhaps at this point, I should clarify my perspective. My daughter majored in computer science and music at a research-oriented university and has had a successful career as a software developer. Her experiences as a woman in computing have been a continuing topic of conversation between us for some fifteen years.

7. Ruth Schwartz Cowan, "The Consumption Junction: A Proposal for Research Strategies in the Sociology of Technology," in *The Social Construction of Technological Systems: New Directions in the Sociology and History of Technology*, ed. Wiebe E. Bijker, Thomas P. Hughes, and Trevor J. Pinch (Cambridge, MA: MIT Press, 1987). Flis Henwood, "Establishing Gender Perspectives on Information Technology: Problems, Issues and Opportunities," in *Gendered by Design? Information Technology and Office Systems*, ed. Eileen Green, Jenny Owen, and Den Pain (London: Taylor and Francis, 1993), chap. 2.

8. "Computer" here means electronic, digital, stored program computer, i.e., a practical device with the capabilities of a Turing machine. One can get bogged down in various definitional problems here, none of which is pertinent to the point at issue. That does not mean anything goes. No one who has worked with computers doubts their capacity to resist (see Andrew Pickering, *The Mangle of Practice*). That resistance goes beyond occasional crashes in the midst of a late-night chapter: it includes planes crashing, rockets going astray and exploding, overdoses of radiation therapy, and collapse of the telephone system. These seem about as real and non-negotiable as the world can get.

9. Judy Wajcman, *Feminism Confronts Technology* (University Park: Pennsylvania State University Press, 1991). Or, to take another tack, counterproductive: "It could equally well be that once these newly-discovered [feminine]

attributes of flexibility, intuition, etc. are revalued and become sought-after skills in computing, men will be the first in line to demonstrate their competence in the field." Henwood, "Establishing," 42–43.

10. T. Estrin, "Women's Studies and Computer Science: Their Intersection," *IEEE Annals of the History of Computing* 18, no. 3 (1996): 43–46. Estrin is Professor Emerita of Computer Science at UCLA and has had a distinguished career in the field of biomedical engineering and computer science, including terms as director of the Engineering, Computer, and Systems Division of the National Science Foundation, 1982–1984, and member of the Board of Directors of the Aerospace Corporation.

11. Estrin, "Women's Studies and Computer Science," 44, 46; Bjarne Stroustrup, *The Design and Evolution of C++* (Reading, MA: Addison-Wesley Publishing Co., 1994). By coincidence Estrin's article is followed in the same special number of *Annals* by Alison Adam's "Constructions of Gender in the History of Artificial Intelligence," in which the line of thinking for which Lisp has served as primary tool, indeed for which it was designed, is designated as irremediably masculinist. It is hard to aim at Lisp without hitting Logo.

12. Alan Kay, "The Early History of Smalltalk," in *History of Programming Languages II*, ed. T. M. Bergin and R. G. Gibson (New York: ACM Press; Reading, MA: Addison-Wesley Publishing Co., 1996), 511–578. Kristen Nygaard and Ole-Johan Dahl, "The Development of the Simula Languages," in *History of Programming Languages*, ed. Richard Wexelblat (New York: Academic Press, 1981), 439–438. M. Douglas McIlroy, "Mass Produced Software Components," in *Software Engineering: Report on a Conference Sponsored by the NATO Science Committee, Garmisch, Germany, 7th to 11th October 1968*, ed. Peter Naur and Brian Randell (Brussels: NATO Scientific Affairs Division, 1969), 138–135; repr. in *Software Engineering: Concepts and Techniques: Proceedings of the NATO Conferences*, ed. Peter Naur, Brian Randell, and J. N. Buxton (New York: Petrocelli/Charter, 1976).

13. Tove Håpnes and Knut Sørensen, "Competition and Collaboration in Male Shaping of Computing: A Study of a Norwegian Hacker Culture," in *The Gender-Technology Relation: Contemporary Theory and Research*, ed. Keith Grint and Rosalind Gill (London: Taylor and Francis, 1995), chap. 7, quote is on p. 177. They refer here to B. Rasmussen and T. Håpnes, "The Production of Male Power in Computer Science," in *Women, Work and Computerization: Understanding and Overcoming Bias in Work and Education—Proceedings of the IFIP TC9/WG 9.1 Conference, Helsinki, Finland, 31 June–2 July 1991*, ed. I. V. Eriksson, B. A. Kitchenham, and K. G. Tijdens (Amsterdam, 1991), and T. Håpnes and B. Rasmussen, "Excluding Women from the Technologies of the Future?" *Futures* (December 1991): 1107–1119.

14. Joseph Weizenbaum, *Computer Power and Human Reason* (San Francisco: W. H. Freeman, 1976), esp. chap. 4, "Science and the Compulsive Programmer"; Sherry Turkle, *The Second Self: Computers and the Human Spirit* (New York: Simon and Schuster, 1984), esp. chap. 3, "Child Programmers: The First Generation." There in discussing styles of programming, Turkle differentiates between hard and soft mastery, which she associates with Claude

Lévi-Strauss's distinction between scientist and *bricoleur*: "Hard mastery is the mastery of the planner, the engineer, soft mastery is the mastery of the artist: try this, wait for a response, try something else, let the overall shape emerge from an interaction with the medium. It is more like a conversation than a monologue" (14–15). Although her first example contrasts the practices of two boys, she goes on to note on p. 19, "But now it is time to state what might be anticipated by many readers: girls tend to be soft masters, while the hard masters are overwhelmingly male." Håpnes and Sørensen, "Competition and Collaboration in Male Shaping of Computing," 189.

15. Peter H. Salus, *A Quarter Century of Unix* (Reading, MA: Addison-Wesley, 1994). For a sense of UNIX as a culture, see Don Libes and Sandy Ressler, *Life with UNIX: A Guide for Everyone* (Englewood Cliffs, NJ: Prentice Hall, 1989).

16. Ulrike Erb, "Exploring the Excluded: A Feminist Approach to Opening New Perspectives in Computer Science," in *Women, Work and Computerization: Spinning a Web from Past to Future* (Proceedings of the 6th International IFIP-Conference, Bonn, Germany, May 24–27, 1997), ed. A. Frances Grundy, Doris Köhler, Veronika Oechtering, and Ulrike Petersen (Berlin: Springer-Verlag, 1997), 201–207; at 203–204.

17. Erb, "Exploring the Excluded," 26.

18. Lucy Suchman, "Supporting Articulation Work: Aspects of a Feminist Practice of Technology Production," in *Women, Work and Computerization: Breaking Old Boundaries—Building New Forms*, ed. Alison Adam, Judy Emms, Eileen Green, and Jenny Owen, 7–21 (Amsterdam: Elsevier, 1994).

19. Frederick P. Brooks, *The Mythical Man-Month: Essays on Software Engineering* (Reading, MA: Addison-Wesley, 1975), 25.

20. Naur and Randell, *Software Engineering*.

21. [Added 2004] On these models, see my "Finding a History for Software Engineering," *IEEE Annals of the History of Computing* 26, no. 1 (2004): 8–19.

22. Perhaps the most prominent is Mary Shaw, professor of Computer Science at Carnegie Mellon University and former chief scientist at the DoD-sponsored Software Engineering Institute there in the late 1980s. Shaw earned her reputation in the area of data abstraction but most recently has emerged as a strong proponent of an architectural approach to software development; see Mary Shaw and David Garlan, *Software Architecture: Perspectives on an Emerging Discipline* (Upper Saddle River, NJ: Prentice Hall, 1996).

23. When I made this point in another context in a talk at the Dibner Institute in 1996, Joel Moses, the provost of MIT, commented that a recent study of their program in software engineering had come to essentially the same conclusion and was revising the curriculum so as to direct students to courses outside computer science.

9. Computing and Mathematics at Princeton in the 1950s

1. For details on the IAS computer and related projects, see William Aspray, *John von Neumann and the Origins of Modern Computing* (Cambridge, MA: MIT Press, 1990).

2. John von Neumann and Oskar Morgenstern, *Theory of Games and Economic Behavior* (Princeton, NJ: Princeton University Press, 1944).

3. Martin Shubik, "Game Theory at Princeton, 1949–1955: A Personal Reminiscence," in *Toward a History of Game Theory*, ed. E. Roy Weintraub (Durham, NC: Duke University Press, 1992): 151–163, at 153 and 161–162.

4. Interview in *The Princeton Mathematics Community in the 1930s: An Oral History Project* (Princeton University, 1985), PMC36, 4.

5. J. McCarthy, M. L. Minsky, N. Rochester, and C. E. Shannon, "A Proposal for the Dartmouth Summer Research Project on Artificial Intelligence," http://www.formal.stanford.edu/jmc/history/dartmouth/dartmouth.html.

6. John von Neumann, "On a Logical and General Theory of Automata," in *Cerebral Mechanisms in Behavior—The Hixon Symposium*, ed. L. A. Jeffries (New York: Wiley, 1951), 1–31; repr. in *Papers of John von Neumann on Computing and Computer Theory*, ed. William Aspray and Arthur Burks (Cambridge, MA: MIT Press; Los Angeles: Tomash Publishers, 1987), 391–431, at 406.

7. Alonzo Church, *Introduction to Mathematical Logic* (Princeton, NJ: Princeton University Press, 1956); Stephen C. Kleene, *Introduction to Metamathematics* (Amsterdam: North-Holland Publishing, 1952).

10. Computer Science: The Search for a Mathematical Theory

1. "An Undergraduate Program in Computer Science—Preliminary Recommendations," *Communications of the ACM* 8, no. 9 (1965): 543–552, at 544.

2. Bruce W. Arden, ed., *What Can Be Automated? The Computer Science and Engineering Research Study (COSERS)* (Cambridge, MA: MIT Press, 1980), 9.

3. Allan Newell, A. J. Perlis, and Herbert Simon, "What Is Computer Science?" [Letter to the editor], *Science* 157 (1967): 1373–1374.

4. Alan Turing, "On Computable Numbers, with an Application to the Entscheidungs Problem," *Proceedings of the London Mathematical Society*, ser. 2, 42 (1936): 230–265, at 231. [Correction, July 2004] The original article spoke here of "three symbols (read/write, shift left/right, next state)," but as an entry in a state table, the description must begin with the current state itself and the input symbol on which the next three actions depend. Each state will have as many entries as there are input symbols.

5. D. Hilbert and W. Ackermann, *Grundzüge der theoretischen Logik* (Berlin: Springer, 1928), 73–74.

6. Stephen C. Kleene, "Origins of Recursive Function Theory," *Annals of the History of Computing* 3, no. 1 (1981): 52–67.

7. Claude E. Shannon, "A Symbolic Analysis of Relay and Switching Circuits," *Transactions of the AIEE* 57 (1938): 713–723.

8. Michael O. Rabin and Dana Scott, "Finite Automata and Their Decision Problems," *IBM Journal* (April 1959): 114–125, at 114.

9. Published in *Cerebral Mechanisms in Behavior—The Hixon Symposium*, ed. L. A. Jeffries (New York: Wiley, 1951), 1–31; repr. in *Papers of John von Neumann on Computing and Computer Theory*, ed. William Aspray and

Arthur Burks (Cambridge, MA: MIT Press; Los Angeles: Tomash Publishers, 1987), 391–431, at 406.

10. Alan M. Turing, "Computing Machinery and Intelligence," *Mind* 236 (October 1950): 433–460. Warren S. McCulloch and Walter Pitts, "A Logical Calculus of the Ideas Immanent in Nervous Activity," *Bulletin of Mathematical Biophysics* 5 (1943): 115–133.

11. "Representation of Events in Nerve Nets and Finite Automata," in *Automata Studies*, ed. Claude Shannon and John McCarthy (Princeton, NJ: Princeton University Press, 1956), 3–41.

12. M. P. Schützenberger, "Une théorie algébrique du codage," *Séminaire P. Dubreil et C. Pisot* (Faculté des Sciences de Paris), Année 1955/56, no. 15 (dated 27 February 1956), 15-02. Cf. "On an Application of Semi Groups[!] Methods to Some Problems in Coding," *IRE Transactions in Information Theory* 2, no. 3 (1956): 47–60.

13. M. P. Schützenberger, "Un probléme de la théorie des automates," *Seminaire Dubreil-Pisol* (1959/60) 13, no. 3 (November 23, 1959): 3-01.

14. Noam Chomsky, "Three Models of Language," *IRE Transactions in Information Theory* 2, no. 3 (1956): 113–124, at 113.

15. Samuel Eilenberg, *Automata, Languages, and Machines*, vol. A (New York: Academic Press, 1974).

16. John McCarthy, "Towards a Mathematical Science of Computation," *Proceedings of the IFIP Congress 62*, ed. Cicely M. Popplewell (Amsterdam: North-Holland Press, 1963), 21–28, at 21.

17. John McCarthy, "A Basis for a Mathematical Theory of Computation," in *Computer Programming and Formal Systems*, ed. P. Braffort and D. Hirschberg (Amsterdam: North-Holland Publishing, 1963), 33–70, at 69.

18. McCarthy, "Mathematical Science."

19. Dana S. Scott, "Outline of a Mathematical Theory of Computation" (Technical Monograph PRG-2, Oxford University Computing Laboratory, 1970), 4.

20. Arden, *What Can Be Automated?*

11. Extracts from *Computers and Mathematics: The Search for a Discipline of Computer Science*

1. Research for this paper was generously supported by the Alfred P. Sloan Foundation through its New Liberal Arts Program. Peter Naur, Brian Randell, and J. N. Buxton, eds., *Software Engineering: Concepts and Techniques: Proceedings of the NATO Conferences* (New York: Petrocelli, 1976), 147.

2. Kenneth Flamm, *Creating the Computer* (Washington, DC: Brookings Institution, 1988). Not all believed that the new science should be mathematical. Several recipients of the ACM's Turing Award addressed the question in their award lectures. Although Marvin Minsky ("Form and Content in Computer Science," 1969) agreed that computers are essentially mathematical machines, he decried the trend toward formalization and urged an experimental, programming approach to understanding them. Allen Newell and Herbert Simon ("Computer Science as Empirical Inquiry: Symbols and Search," 1975)

took an even stronger empirical stand, arguing that computer science is the science of computers and that the limits and possibilities of computing could be determined only through experience in using them. Donald E. Knuth ("Computer Programming as an Art," 1974) argued that programming was irreducibly a craft skill, which would resist the automation implicit in a mathematization of computer science. See *ACM Turing Award Lectures: The First Twenty Years, 1966–1985* (New York: ACM Press, 1987). The new subject comprised fields taken from various headings. Programming theory, algorithms, symbolic computation, and computational complexity and efficiency had been the province of numerical analysis. From "Information and Communication" came automata theory, linguistics and formal languages, and information retrieval. To these established categories were added adaptive systems, theorem proving, artificial intelligence and pattern recognition, and simulation.

3. For a fuller sketch and further reading, see M. S. Mahoney, "The History of Computing in the History of Technology," *Annals of the History of Computing* (hereafter *AHC*) 10 (1988): 113–125, and "Cybernetics and Information Technology," in *Companion to the History of Modern Science*, ed. R. C. Olby, G. N. Cantor, J. R. R. Christie and M. J. S. Hodge (London: Routledge, Chapman and Hall, 1989), chap. 34. W. S. McCulloch and W. Pitts, "A Logical Calculus of the Ideas Imminent in Nervous Activity," *Bulletin of Mathematical Biophysics* 5 (1943): 115–133. J. von Neumann, "The General and Logical Theory of Automata," in *Cerebral Mechanisms in Behavior: The Hixon Symposium*, ed. L. A. Jeffries (New York: Wiley, 1951). Robert McNaughton, "The Theory of Automata, a Survey," *Advances in Computing* 2 (1961): 379–421.

4. See, for example, J. Hartmanis and R. E. Stearns, *Algebraic Structure of Sequential Machines* (Englewood Cliffs, NJ: Prentice Hall, 1966), and Paul M. Cohn, *Universal Algebra* (Dordrecht: Reidel, 1965; 2nd rev. ed., 1981). Cf. Alfred North Whitehead, *A Treatise of Universal Algebra* (Cambridge, 1898), I, 29: "[Boole's algebra, characterized by the relation $a = a + a$,] leads to the simplest and most rudimentary type of algebraic symbolism. No symbols representing number or quantity are required in it. The interpretation of such an algebra may be expected therefore to lead to an equally simple and fundamental science. It will be found that the only species of this genus which at present has been developed is the Algebra of Symbolic Logic, though there seems no reason why other algebras of this genus should not be developed to receive interpretations in fields of science where strict demonstrative reasoning without relation to number or quantity is required." M. O. Rabin, "Speed of Computation and Classification of Recursive Sets," *Third Convention of Scientific Societies* (Israel, 1959), 1–2; "Degree of Difficulty of Computing a Function and a Partial Ordering of Recursive Sets," *Technical Report No. 2, ONR Contract* (Hebrew University Jerusalem, 1960). J. Hartmanis and R. E. Stearns, "On the Computational Complexity of Algorithms," *Transactions of the AMS* 117 (1965): 285–306. S. Cook, "The Complexity of Theorem Proving Procedures," *Proceedings of the Third Annual ACM Symposium on Theory of Computing (STOC)* (New York: Association for Computing

Machinery, 1971), 151–158. R. Karp, "Reducibility among Combinatorial Problems," in *Complexity of Computer Computations* (New York: Plenum, 1972), 85–104.

5. S. C. Kleene, "Representation of Events in Nerve Nets and Finite Automata," in *Automata Studies*, ed. J. McCarthy and C. E. Shannon (Princeton: Princeton University Press, 1956), 3–41. M. O. Rabin and D. S. Scott, "Finite Automata and Their Decision Problems," *IBM Journal of Research and Development* 3 (April 1959): 114–124. Sheila A. Greibach, "Formal Languages: Origins and Directions," *AHC* 3, no. 1 (1981): 14–41.

6. For example, John McCarthy, in an article to be discussed below, argued that none of the three current (1961) directions of research into the mathematics of computing held much promise of such a science. Numerical analysis was too narrowly focused. The theory of computability set a framework into which any mathematics of computation would have to fit, but it focused on what was unsolvable rather than seeking positive results, and its level of description was too general to capture actual algorithms. Finally, the theory of finite automata, though it operated at the right level of generality, exploded in complexity with the size of current computers. As he explained in another article, "The fact of finiteness is used to show that the automaton will eventually repeat a state. However, anyone who waits for an IBM 7090 to repeat a state, solely because it is a finite automaton, is in for a very long wait." ("Towards a Mathematical Science of Computation," *Information Processing 1962: Proceedings of IFIP Congress 62* (Amsterdam: North-Holland, 1963). "Attaching Meaning to Programs," in *Mathematical Aspects of Computer Science* (Proceedings of Symposia in Applied Mathematics, 19; Providence: AMS, 1967), 19–32.

7. J. Barkley Rosser, "Highlights of the History of the Lambda Calculus," *AHC* 6 (1984): 337–349. S. C. Kleene, "Origins of Recursive Function Theory," *AHC* 3 (1981): 52–67.

8. Reprinted, with corrections and an added tenth section, in *Computer Programming and Formal Systems*, ed. P. Braffort and D. Hirschberg (Amsterdam: North-Holland, 1963), 33–70.

9. "Recursive Functions of Symbolic Expressions and Their Computation by Machine," *Communications of the ACM* 3, no. 4 (1960): 184–195. Interview, 3 December 1990. P. J. Landin, "The Mechanical Evaluation of Expressions," *Computer Journal* 6 (1964): 308–320.

10. McCarthy and his co-workers had encountered this problem in designing LISP; it came to be called the FUNARG problem. See his account in *History of Programming Languages*, ed. R. Wexelblat (New York: Academic Press, 1981). The right side of the equation is a conditional expression, which consists of a list of conditional propositions to be evaluated in order from left to right and which takes the value of the consequent of the first proposition of which the antecedent is true. In the above expression, if $n = 0$, the value is m, otherwise (T is always true) it is $g(m',n)$; for example, $g(3,2) = g(4,1) = g(5,0) = 5$. More precisely, in McCarthy's system, they satisfy the relation $f = label(f, \lambda m.\lambda n.(n = 0 \rightarrow m', T \rightarrow f(m',n)))$.

11. P. J. Landin, "The Mechanical Evaluation of Expressions," *Computer Journal* 6 (1964): 308–320, develops a "syntactically sugared," λ-less version of Church's notation, which Landin later used to set out a formal specification of the semantics of ALGOL 60. Others undertook to take the approach into the realm of semigroups and categories. In *Mathematical Reviews* 26 (1963), #5766, Nerode wrote that McCarthy had introduced "yet another definition of computability" via conditional expressions and recursive induction. The former is "an arithmetical convenience for handling definition by cases," and the latter, on which McCarthy laid great stress, "is nothing else but the uniqueness of the object defined by a recursive definition." "In the reviewer's opinion," he concluded, "the problem of justifying the title is still open."

12. D. S. Scott, "Logic and Programming Languages" (1976 Turing Award Lecture), in *ACM Turing Award Lectures*, 47–62. Dana S. Scott, "Outline of a Mathematical Theory of Computation," *Proceedings of the Fourth Annual Princeton Conference on Information Sciences and Systems* (1970); rev. and expanded as Technical Monograph PRG-2, Oxford University Computing Laboratory, 1970, 2–3.

13. Ibid., 4–5.

14. Technical Monograph PRG-6, Oxford University Computing Laboratory, 1971, 40; also published in *Proceedings of the Symposium on Computers and Automata*, Microwave Research Institute Symposia Series, vol. 21, Polytechnic Institute of Brooklyn, 1971.

12. The Structures of Computation and the Mathematical Structure of Nature

1. Birkhoff modestly omits mention here of his textbook, *A Survey of Modern Algebra*, written jointly with Saunders MacLane, which played a major role in that development. How and why the shift came about is, of course, another question of considerable historical interest; see Leo Corry, *Modern Algebra and the Rise of Mathematical Structures* (Basel: Birkhäuser, 1996; 2nd. ed., 2004).

2. Garrett Birkhoff and Thomas C. Bartee, *Modern Applied Algebra* (New York: McGraw-Hill Book Company, 1970), preface, v. A preliminary edition appeared in 1967.

3. On the computer as a "defining technology," see J. David Bolter, *Turing's Man* (Chapel Hill: University of North Carolina Press, 1984).

4. "Software as Science—Science as Software," in *History of Computing: Software Issues*, ed. Ulf Hashagen, Reinhard Keil-Slawik, and Arthur Norberg (Berlin: Springer-Verlag, 2002), 25–48.

5. See, in addition to the usual suspects, the illuminating argument for a material epistemology of science by Davis Baird, *Thing Knowledge* (Berkeley: University of California Press, 2004). Paul Humphreys offers a counterpart for computational science in *Extending Ourselves: Computational Science, Empiricism, and Scientific Method* (Oxford: Oxford University Press, 2004). Francis Bacon, *The New Organon* [London, 1620], trans. James Spedding,

Robert L. Ellis, and Douglas D. Heath in the *The Works*, vol. 8 (Boston: Taggart and Thompson, 1863; repr., Indianapolis: Bobbs-Merrill, 1960) Aph. 3.

6. On the last, see Ehud Shapiro's contribution to the conference, "What Is a Computer? The World Is Not Enough to Answer," describing computers made of biological molecules. In recent articles and interviews arguing that computer science should rank on a par with the natural sciences, Peter Denning points to recent "discoveries" that nature is based on information processes. See his "Great Principles of Computing" Web site. Using the word "discovery" places those processes in the world rather than in our representations of it. In a report prepared for a panel discussion on graduate education in computer science at the National ACM Meeting in September 1970, Peter Wegner stood on firmer philosophical ground in predicting what it would mean to view the world in computational terms: "Computer science may affect the conceptual framework of science, philosophy and epistemology as fundamentally as physics or mathematics. It may revolutionize our conceptual environment just as fundamentally as computer technology has revolutionized our physical environment. This conceptual revolution will come about not by the once-popular but increasingly discredited notion that the 'artificial intelligence' of computers will replace the 'natural intelligence' of humans, but rather by the introduction into the fabric of our thinking of principles which govern computational transformations of information structures." ("Some Thoughts on Graduate Education in Computer Science," *ACM SIGCSE Bulletin* 2, no. 4 [1970]: 34).

7. Murray Gell-Mann and Yuval Ne'eman, *The Eightfold Way* (New York: W. A. Benjamin, 1964). Eugene P. Wigner, "The Unreasonable Effectiveness of Mathematics in the Natural Sciences," *Communications in Pure and Applied Mathematics* 13, no. 1 (1960). Cf. Mark Steiner, *The Applicability of Mathematics as a Philosophical Problem* (Cambridge, MA: Harvard University Press, 1998).

8. On the seventeenth-century origins of the structural notion of algebra, see Michael S. Mahoney, "The Beginnings of Algebraic Thought in the Seventeenth Century," in *Descartes: Philosophy, Mathematics and Physics*, ed. S. Gaukroger (Sussex: Harvester Press; Totowa, NJ: Barnes and Noble Books, 1980), chap. 5. For subsequent developments, see the book by Corry cited above. Herbert Mehrtens, *Moderne Sprache Mathematik: Eine Geschichte des Streits um die Grundlagen der Disziplin und des Subjekts formaler Systeme* (Frankfurt am Main: Suhrkamp, 1990). For a more nuanced analysis of Hilbert's thinking on the relation of mathematics to the physical world, see Leo Corry, *David Hilbert and the Axiomatization of Physics (1898–1918): From Grundlagen der Geometrie to Grundlagen der Physik* (Dordrecht: Kluwer Academic Publishers, 2004.

9. Bruna Ingrao and Giorgio Israel, *The Invisible Hand: Economic Equilibrium in the History of Science*, trans. Ian McGilvray (Cambridge, MA: MIT Press, 1990), 186–187.

10. M. P. Schützenberger, "A Propos de la 'Cybernétique,'" *Evolution Psychiatrique* 20, no. 4 (1949): 595, 598.

11. For an overview of von Neumann's work in the first area, see William Aspray, *John von Neumann and the Origins of Modern Computing* (Cambridge, MA: MIT Press, 1990), chap. 5, "The Transformation of Numerical Analysis."

12. John von Neumann, "The General and Logical Theory of Automata," in *Cerebral Mechanisms in Behavior—The Hixon Symposium*, ed. L. A. Jeffries (New York: Wiley, 1951), 1–31; repr. in *Papers of John von Neumann on Computing and Computer Theory*, ed. William Aspray and Arthur Burks (Cambridge, MA: MIT Press, 1987), 391–431.

13. See in particular "Computer Science: The Search for a Mathematical Theory," in *Science in the 20th Century*, ed. John Krige and Dominique Pestre (Amsterdam: Harwood Academic Publishers, 1997), chap. 31, and "What Was the Question? The Origins of the Theory of Computation," in *Using History to Teach Computer Science and Related Disciplines* (Selected Papers from a Workshop Sponsored by Computing Research Association with Funding from the National Science Foundation), ed. Atsushi Akera and William Aspray (Washington, DC: Computing Research Association, 2004), 225–232.

14. For all the public fervor surrounding Andrew Wiles's proof of Fermat's "last theorem," it was the new areas of investigation suggested by his solution of the Taniyama conjecture that excited his fellow practitioners.

15. W. S. McCulloch and W. Pitts, "A Logical Calculus of the Ideas Immanent in Nervous Activity," *Bulletin of Mathematical Biophysics* 5 (1943): 115–133; D. A. Huffman, "The Synthesis of Sequential Switching Circuits," *Journal of the Franklin Institute* 257, no. 3 (1954): 161–190. E. F. Moore, "Gedanken-Experiments on Sequential Machines," and Stephen C. Kleene, "The Representation of Events in Nerve Nets and Finite Automata," in *Automata Studies*, ed. John McCarthy and Claude E. Shannon (Princeton, NJ: Princeton University Press, 1956), 129–153, 3–41. Kleene's article derived from a RAND Memorandum (RM-704) with the same title, dated 15 December 1951. Michael Rabin and Dana Scott, "Finite Automata and Their Decision Problems," *IBM Journal of Research and Development* 3, no. 2 (April 1959): 114–125.

16. Kleene, "Representation of Events," 37. George A. Miller dates the beginnings of cognitive science from this symposium, which included a presentation by Allen Newell and Herbert Simon of their "Logic Theory Machine," Noam Chomsky's "Three Models for the Description of Language," and Miller's own "Human Memory and the Storage of Information." Ironically, for Chomsky and Miller, the symposium marked a turn away from information theory as the model for their work. M. P. Schützenberger, "Un problème de la théorie des automates," *Seminaire Dubreil-Pisot* 13 (1959–1960): 3 (23 November 1959); "On the Definition of a Family of Automata," *Information and Control* 4 (1961): 245–270; "Certain Elementary Families of Automata," in *Mathematical Theory of Automata*, ed. Jerome Fox (Brooklyn, NY: Polytechnic Press, 1963), 139–153.

17. In 1955, Chomsky began circulating the manuscript of his *Logical Structure of Linguistic Theory*, but MIT's Technology Press hesitated to publish such a radically new approach to the field before Chomsky had exposed it to critical

examination through articles in professional journals. A pared-down version
of the work appeared as *Syntactic Structures* in 1957, and a partially revised
version of the whole, with a retrospective introduction, in 1975 (New York:
Plenum Press). Noam Chomsky, "Three Models of Language," *IRE Transactions in Information Theory* 2, no. 3 (1956): 113–124, at 113.

18. Noam Chomsky, "On Certain Formal Properties of Grammars," *Information and Control* 2, no. 2 (1959): 137–167. Paul C. Rosenbloom, *The Elements of Mathematical Logic* (New York: Dover Publications, 1950; repr., 2005). In a review in the *Journal of Symbolic Logic* 18 (1953): 277–280, Martin Davis drew particular attention to Rosenbloom's inclusion of combinatory logics and Post canonical systems, joining the author in hoping the book would make Post's work more widely known. To judge from citations, Rosenbloom's text was an important resource for many of the people discussed here. For example, it appears that Peter Landin and Dana Scott first learned of Church's λ-calculus through it. M. P. Schützenberger, "Some Remarks on Chomsky's Context-Free Languages" (Quarterly Progress Report, (MIT) Research Laboratory for Electronics 63, October 15, 1961), 155–170. Seymour Ginsburg and H. Gordon Rice, "Two Families of Languages related to ALGOL," *Journal of the ACM* 9 (1962): 350–371. J. W. Backus, "The Syntax and Semantics of the Proposed International Algebraic Language of the Zurich ACM-GAMM Conference," *Proceedings of the International Conference on Information Processing* (Paris: UNESCO, 1959), 125–132. Backus later pointed to Post's production systems as the model for what he referred to here as *metalinguistic formulae.* Alfred Tarski, "A Lattice-Theoretical Fixpoint Theorem and Its Applications," *Pacific Journal of Mathematics* 5 (1955): 285–309. Klaus Samelson and Friedrich Bauer, "Sequentielle Formelübersetzung," *Elektronische Rechenanlagen* 1 (1959): 176–182; "Sequential Formula Translation," *Communications of the ACM* 3, no. 2 (1960): 76–83. Noam Chomsky, "Context-Free Grammars and Pushdown Storage" (Quarterly Progress Report, (MIT) Research Laboratory for Electronics 65, April 15, 1962), 187–194. Noam Chomsky and M. P. Schützenberger, "The Algebraic Theory of Context-Free Languages," in *Computer Programming and Formal Systems,* ed. P. Braffort and D. Hirschberg (Amsterdam: North-Holland Publishing Company, 1963). Originally prepared for an IBM Seminar in Blaricum, Netherlands, in the summer of '61, the published version took account of intervening results, in particular the introduction of the pushdown store.

19. Dominique Perrin, "Les débuts de la théorie des automates," *Technique et Science Informatique* 14 (1995): 409–443. Michael O. Rabin, "Lectures on Classical and Probabilistic Automata," in *Automata Theory,* ed. E. A. Caianiello (New York: Academic Press, 1966), 306.

20. John McCarthy, "Recursive Functions of Symbolic Expressions and Their Computation by Machine, Part 1," *Communications of the ACM* 3, no. 4 (1960): 184–195; "A Basis for a Mathematical Theory of Computation," *Proceedings of the Western Joint Computer Conference* (New York: NJCC, 1961), 225–238; republished with an addendum "On the Relations between Computation and Mathematical Logic" in *Computer Programming and Formal Sys-*

tems, ed. P. Braffort and D. Hirschberg (Amsterdam: North-Holland, 1963), 33–70; "Towards a Mathematical Science of Computation," *Proceedings IFIP Congress 62*, ed. C. M. Popplewell (Amsterdam: North-Holland, 1963), 21–28. McCarthy cited Church's *Calculi of Lambda Conversion* (Princeton, 1941) but did not name Curry nor give a source for his combinatory logic.

21. McCarthy, "Basis for a Mathematical Theory," 225.

22. McCarthy, "Towards a Mathematical Science," 21.

23. As Anil Nerode pointed out in *Mathematical Reviews* 26 (1963): #5766.

24. P. J. Landin, "A λ-Calculus Approach," in *Advances in Programming and Non-Numerical Computing*, ed. L. Fox (Oxford: Pergamon Press, 1966), chap. 5, p. 97. The main body of the lectures had already appeared as "The Mechanical Evaluation of Expressions," *Computer Journal* 6 (1964): 308–320. In "The Next 700 Programming Languages," *Communications of the ACM* 9, no. 3 (1966): 157–166, Landin expanded the notation into a "family of languages" called *ISWIM* (If You See What I Mean), which he noted could be called "Church without Lambda." On CPL, see D. W. Barron, J. N. Buxton, D. F. Hartley, E. Nixon, and C. Strachey "The Main Features of CPL," *Computer Journal* 6 (1964): 134–142.

25. Christopher Strachey, "Towards a Formal Semantics," in *Formal Language Description Languages for Computer Programming*, ed. T. B. Steel, Jr. (Amsterdam: North-Holland, 1966), 198–216. That is, at each step of a computation, σ is the set of pairs (α, β) relating L-values to R-values, or the current map of the store.

26. Dana Scott, "Outline of a Mathematical Theory of Computation," Technical Monograph PRG-2, Oxford University Computing Laboratory, November 1970, 4; an earlier version appeared in the Proceedings of the Fourth Annual Princeton Conference on Information Sciences and Systems, 1970, 169–176. See also his "Continuous Lattices," in *Toposes, Algebraic Geometry and Logic*, ed. F. W. Lawvere, Springer Lecture Notes in Mathematics, 274 (Berlin: Springer-Verlag, 1972), 97–136. In "A Type-Theoretical Alternative to ISWIM, CUCH, OWHY," a privately circulated paper written in October 1969 and not published until 1993 (in *Theoretical Computer Science* 121:411–440), Scott wrote "Now, it may turn out that a system such as the λ-calculus will have an interpretation along standard lines (and I have spent more days than I care to remember trying to find one), but until it is produced I would like to argue that its purposes can just as well be filled by a system involving types." He found the interpretation a month later.

27. "Semantics of Assignment," *Machine Intelligence* 2 (1968): 3–20, at 3.

28. "Outline of a Mathematical Theory of Computation," Princeton version, 169.

29. In "Recursive Functions," McCarthy used an abstract LISP machine, leaving the question of its implementation to a concluding and independent section.

30. R. M. Burstall and P. J. Landin, "Programs and Their Proofs, an Algebraic Approach," *Machine Intelligence* 4 (1969): 17.

31. Cf. J. Richard Büchi, "Algebraic Theory of Feedback in Discrete Systems, Part I," in *Automata Theory*, ed. E. R. Caianello (New York: Academic

Press, 1966), 71. The reference is to Birkhoff's *Lattice Theory* (New York, 1948).

32. Burstall and Landin, "Programs and Their Proofs," 32 (redrawn).

33. Rod Burstall, "Electronic Category Theory," *Mathematical Foundations of Computer Science 1980* (Proceedings of the Ninth International Symposium—Springer LNCS 88) (Berlin: Springer-Verlag, 1980), 22–39.

34. Paul M. Cohn, *Universal Algebra*, 2nd ed. (Dordrecht: Reidel, 1981), 345. The chapter was reprinted from the *Bulletin of the London Mathematical Society* 7 (1975): 1–29. Looking back on the development of the field, Gian-Carlo Rota wrote in *Indiscrete Thoughts* (Boston: Birkhäuser Verlag, 1997), 221, "Universal algebra has made it. Not, as the founders wanted, as the unified language for algebra, but rather, as the proper language for the unforeseen and fascinating algebraic systems that are being discovered in computer algebra, like the fauna of a new continent." Garrett Birkhoff, "The Role of Modern Algebra in Computing," *Computers in Algebra in Number Theory* (Providence, RI: American Mathematical Society, 1971), 1–47, repr. in his *Selected Papers on Algebra and Topology* (Boston: Birkhäuser, 1987), 513–559, at 517; emphasis in the original.

35. *Automata, Languages, and Machines*, 2 vols. (New York: Columbia University Press, 1974), vol. A, xiii.

36. Aristid Lindenmayer, "Mathematical Models for Cellular Interactions in Development," *Journal of Theoretical Biology* 18 (1968). For subsequent development of the field, see Lindenmayer and Grzegorz Rozenberg, eds., *Automata, Languages, and Development* (Amsterdam: North-Holland, 1976), and Rozenberg and Arto Salomaa, *The Book of L* (Berlin: Springer-Verlag, 1986). On the λ-calculus, see, e.g., W. Fontana and L. W. Buss, "The Barrier of Objects: From Dynamical Systems to Bounded Organizations," in *Boundaries and Barriers*, ed. J. Casti and A. Karlqvist (New York: Addison-Wesley, 1996), 56–116; on the ð-calculus, see A. Regev, W. Silverman, and E. Shapiro, "Representation and Simulation of Biochemical Processes Using the ð-calculus Process Algebra," *Pacific Symposium on Biocomputing* 6 (2001): 459–470, and Regev and Shapiro, "Cellular Abstractions: Cells as Computation," *Nature* 419 (2002): 343. The lengthy quotation is from Corrado Priama, preface to *Computational Methods in Systems Biology*, Proceeding of the First International Workshop, Rovereto, Italy, 24–26 February 2003, LNCS 2602 (Berlin: Springer, 2003).

37. Paul Humphreys, *Extending Ourselves: Computational Science, Empiricism, and Scientific Method* (Oxford: Oxford University Press, 2004), 8.

38. McCarthy, "Basis," 69.

39. Stephen Wolfram, "Computation Theory of Cellular Automata," *Communications in Mathematical Physics* 96 (1984): 15–57, at 16. Repr. in his *Theory and Applications of Cellular Automata* (Singapore: World Scientific, 1986), 189–231, and his *Cellular Automata and Complexity: Collected Papers* (Reading: Addison Wesley, 1994), 159–202.

40. Christopher G. Langton, "Artificial Life" (1989) in *The Philosophy of Artificial Life*, ed. Margaret A. Boden (Oxford: Oxford University Press, 1996), 47.

41. John H. Holland, *Hidden Order: How Adaptation Builds Complexity* (Reading, MA: Addison-Wesley, 1995), 161–162.

42. Ibid., 171–172. Schützenberger, "A Propos de la 'Cybernétique,' " 598.

13. Extracts from *Software as Science—Science as Software*

1. Herbert Simon, *The Sciences of the Artificial* (Cambridge, MA: MIT Press, 1969; 2nd ed., 1981; 3rd ed., 1996). Let me leave aside for the moment questions about how much "nature" ever presents itself to us directly.

2. Marvin Minsky, "Form and Content in Computer Science" (1969 Turing Award), *ACM Turing Award Lectures: The First Twenty Years, 1966–1985* (New York: ACM Press, 1987), 219–242. Allan Newell and Herbert Simon, "Computer Science as Empirical Inquiry: Symbols and Search" (1975 Turing Award), *ACM Turing Award Lectures,* 287–313. Newell and Simon had earlier joined with Alan Perlis in taking a similar position in "What Is Computer Science?" a letter to the editor of *Science* 157 (22 September 1967), 1373–1374. Peter Wegner, "Three Computer Cultures: Computer Technology, Computer Mathematics, and Computer Science," *Advances in Computers* 10 (1970), 7–78. Forsythe, "What to Do until the Computer Scientist Comes," in *American Mathematical Monthly* 75, no. 5 (1968): 454–462; Pierce, in Keynote Address, Conference on Academic and Related Research Programs in Computing Science, 5–8 June 1967; publ. in University Education in Computing Science, ed. Aaron Finerman (New York: Academic Press, 1968), 7, 24.

3. The Turing Award is considered the ACM's highest honor. "It is given to an individual selected for contributions of a technical nature made to the computing community. The contributions should be of lasting and major technical importance to the computer field" (http://www.acm.org/awards/taward . html). A look at the list shows that "technical" has usually (but not always) been construed as "theoretical," indeed "mathematical."

4. It is curious that to this day the community distinguishes between computer science and *theoretical* computer science, as if the former involves some kind of science other than theoretical science. It is not clear what that other kind of science might be nor what is scientific about it.

5. Michael S. Mahoney, "Computer Science: The Search for a Mathematical Theory," in *Science in the 20th Century,* ed. John Krige and Dominique Pestre (Amsterdam: Harwood Academic Publishers, 1997), Chap. 31; see also "The Structures of Computation," in *The First Computers—Histories and Architectures,* ed. Raul Rojas and Ulf Hashagen (Cambridge, MA: MIT Press, 2000).

6. For this and other examples of feedback from computer science to mathematics, see Garrett Birkhoff, "The Role of Modern Algebra in Computing," *Computers in Algebra in Number Theory* (Providence, RI: American Mathematical Society, 1971), 1–47. For a discussion of the changes in the mathematics curriculum prompted by computer science, see Anthony Ralston, "Computer Science, Mathematics, and the Undergraduate Curriculum in Both," *American Mathematical Monthly* 81, no. 7 (1981): 472–485.

7. See, for example, Garrett Birkhoff and Thomas C. Bartee, *Modern Applied Algebra* (New York: McGraw-Hill Book Company, 1970); Rudolf Lidl and Gunter Pilz, *Applied Abstract Algebra* (New York: Springer-Verlag, 1984); Andrea Asperti, *Categories, Types, and Structures: An Introduction to Category Theory for the Working Computer Scientist* (Cambridge, MA: MIT Press, 1991). See W. Fontana and Leo W. Buss, "The Barrier of Objects: From Dynamical Systems to Bounded Organizations," in *Boundaries and Barriers,* ed. J. Casti and A. Karlqvist (Reading, MA: Addison-Wesley, 1996), 56–116.

8. The proceedings of the conference were published in *Communications of the ACM* 7, no. 2 (1964): 51–136; see in particular the "Summary Remarks" by Saul Gorn and the "General Discussion" that followed, 133–136.

9. Such differences shine through the protocols of the Software Engineering Conferences at Garmisch and Rome. People differed about what it would mean to make the subject scientific, about the extent to which one can do so, about the importance of trying to make it so, about the means for achieving that goal. They had different agendas. Not all agendas have converged on the current configuration. For example, Ershov and other Russian computer scientists took their own approach to a science of software but did so in relative isolation from research in the West. To the historian, this independent line of development offers an opportunity for comparisons and contrasts, and holds out the possibility of linking agendas to the political, social, and economic context within which they take shape. See, for example, Andrei P. Ershov, *Origins of Programming: Discourses on Methodology* (New York: Springer-Verlag, 1990); Ershov and M. R. Shura-Bura, "The Early Development of Programming in the USSR," in *A History of Computing in the Twentieth Century,* ed. N. Metropolis, J. Howlett and Gian-Carlo Rota(New York: Academic Press, 1978), 137–196; R. A. Di Paola, "A Survey of Soviet Work in the Theory of Computer Programming" (Rand Memorandum RM-5424-PR, Rand Corporation, Santa Monica, 1967).

10. Arthur L. Norberg and Judy E. O'Neill, *Transforming Computer Technology: Information Processing for the Pentagon, 1962–1986* (Baltimore: Johns Hopkins University Press, 1996). William Aspray, Bernard O. Williams, and Andrew H. Goldstein, "Computing as Servant and Science: The Impact of the National Science Foundation" (unpublished draft, 1992); cf. Aspray and Williams, "Arming American Scientists: NSF and the Provision of Scientific Computing Facilities for Universities, 1950–1973," *Annals of the History of Computing* 16, no. 4 (1994): 60–74. The lengthy quotation is from Richard W. Hamming, "One Man's View of Computer Science," *ACM Turing Award Lectures,* 207–218, at 208. [Hamming's note]: G. E. Forsythe, "What to Do until the Computer Scientist Comes," *American Mathematical Monthly* 75, no. 5 (May 1968): 454–461.

11. Available at http://www-formal.stanford.edu/jmc/history/dartmouth.html.

12. Thomas S. Kuhn, *The Structure of Scientific Revolutions* (Chicago: University of Chicago Press, 1962; 2nd ed., 1970).

13. For example, ACM Curriculum Committee on Computer Science, "An Undergraduate Program in Computer Science—Preliminary Recommendations,"

Communications of the ACM 8 (1965): 543–548; "Curriculum 68—Recommendations for Academic Programs in Computer Science," *Communications of the ACM* 11 (1968): 151–197; "Curriculum 78—Recommendations for the Undergraduate Program in Computer Science," *Communications of the ACM* 22 (1979): 147–166; A. Ralston and M. Shaw, "Curriculum 78—Is Computer Science Really That Unmathematical?" *Communications of the ACM* 23 (1980): 67–70.

14. For Purdue, see John Rice and Richard A. DeMillo, eds., *Studies in Computer Science in Honor of Samuel D. Conte* (New York: Plenum Press, 1994); for Cornell, David Gries, "Twenty Years of Computer Science at Cornell," *Engineering: Cornell Quarterly* 20, no. 2 (1985): 2–11.

15. Peter Naur, Brian Randell, and J. N. Buxton, eds., *Software Engineering: Concepts and Techniques. Proceedings of the NATO Conferences* (New York: Petrocelli, 1976), 147. The COSERS study offered three main three main reasons for its assertion: "(1) Computers and programs are inherently mathematical objects. They manipulate formal symbols, and their input-output behavior can be described by mathematical functions. The notations we use to represent them strongly resemble the formal notations which are used throughout mathematics and systematically studied in mathematical logic. (2) Programs often accept arbitrarily large amounts of input data; hence, they have a potentially unbounded number of possible inputs. Thus a program embraces, in finite terms, an infinite number of possible computations; and mathematics provides powerful tools for reasoning about infinite numbers of cases. (3) Solving complex information-processing problems requires mathematical analysis. While some of this analysis is highly problem-dependent and belongs to specific application areas, some constructions and proof methods are broadly applicable, and thus become the subject of theoretical computer science. Bruce W. Arden, ed., *What Can Be Automated?* (Cambridge, MA: MIT Press, 1980), 139. The committee consisted of Richard M. Karp (chair; Berkeley), Zohar Manna (Stanford), Albert R. Meyer (MIT), John C. Reynolds (Syracuse), Robert W. Ritchie (Washington), Jeffrey D. Ullman (Stanford), and Shmuel Winograd (IBM Research).

16. C. A. R. Hoare, "The Mathematics of Programming," in his Essays in Computing Science (Hemel Hempstead, 1989), 352.

17. For a discussion of some of the issues, together with case studies, see Andrew Pickering, *The Mangle of Practice: Time, Agency, and Science* (Chicago: University of Chicago Press, 1995).

18. Arden, ed., *What Can Be Automated?* 9?

19. Frederick P. Brooks, "No Silver Bullet—Essence and Accidents of Software Engineering," *Information Processing 86*, ed. H. J. Kugler (Amsterdam: Elsevier Science, 1986), 1069–1076; repr. in *Computer* 20, no. 4 (1987): 10–19; and in the Anniversary Edition of *The Mythical Man-Month: Essays on Software Engineering* (Reading, MA: Addison-Wesley, 1995), chap. 16. Chapter 17, "'No Silver Bullet' Refired" is a response to critics of the original article and a review of the silver bullets that have missed the mark over the

intervening decade. Since the late 1980s, this interaction has been moved to a meta-level, as researchers have sought to model the process by which the computational model is designed. See, for example, Leon Osterweil's "Software Processes Are Software Too," *Proceedings: 9th International Conference on Software Engineering* (Los Angeles, CA: IEEE Computer Society Press, 1987), 2–13.

20. In introducing the ACM to readers of the first issue of its *Journal* in January 1954, S. B. Williams anticipated the coinage: "Until the engineering societies, AIEE and IRE, became sufficiently interested to struggle with 'hardware,' the Association provided a forum for all phases of the field. Now the Association can direct its efforts to the other phases of computing systems, such as numerical analysis, logical design, application and use, and, last but not least, to programming." (*Journal of the ACM* 1, no. 1 [1954]: 3). John Tukey, "The Teaching of Concrete Mathematics," *American Mathematical Monthly* 65, no. 1 (1958): 1–9, at 2: "Today the 'software' comprising the carefully planned interpretive routines, compilers, and other aspects of automative [*sic*] programming are at least as important to the modern electronic calculator as its 'hardware' of tubes, transistors, wires, tapes and the like."

21. Jan van Leeuwen, ed., *Handbook of Theoretical Computer Science* (Amsterdam: Elsevier; Cambridge, MA: MIT Press, 1990), vol. A, *Algorithms and Complexity*; vol. B, *Formal Models and Semantics*.

22. Only after writing these words did I discover that Wolfgang Coy had asked the same question, "'Informatique': What's in a Name?" and had reached the same conclusion in his essay "Defining Discipline," in *Foundations of Computer Science: Potential—Theory—Cognition*, ed. Ch. Freksa, M. Jantzen, R. Valk (Berlin: Springer, 1997): "But the German 'Informatik' made a strange twist: While it uses the French word, it sticks firmly to the American usage of computer science (with elements from computer engineering)." Together with many colleagues, Coy would like Informatik to be something quite different; but it is not yet, and it has not been so historically. See, inter alia, his "Für eine Theorie der Informatik!" in *Sichtweisen der Informatik*, ed. Wolfgang Coy, Frieder Nake, and Jörg-Martin Pflüger (Braunschweig: Vieweg, 1992), 17–32.

23. See Robert Rosen, "Effective Processes and Natural Law," in *The Universal Turing Machine: A Half-Century Survey*, ed. Rolf Herken (Vienna: Springer-Verlag, 1994–1995), 485–498; cf. his earlier article, "Church's Thesis and Its Relation to the Concept of Realizability in Biology and Physics," *Bulletin of Mathematical Biophysics* 24 (1962): 375–393. [Added 2005] Stephen Wolfram has since made an extensive case for this proposition in *A New Kind of Science* (Champaign, IL: Wolfram Media, 2002).

Éloge

1. M. S. Mahoney, *The Mathematical Career of Pierre de Fermat* (Princeton, NJ: Princeton University Press, 1973), x.

2. Ibid.

3. M. S. Mahoney, "The Mathematical Realm of Nature," in *The Cambridge History of Seventeenth Century Philosophy*, ed. D. Garber and M. Ayers (Cambridge: Cambridge University Press, 1998) vol. 1, 744–745.

4. Cited from the oral version of the presentation, "Histories of Computing(s)," delivered 20 March 2004 as part of the *Digital Scholarship, Digital Culture* series, and available at http://www.princeton.edu/~hos/Mahoney/articles/histories/kingscch.htm (accessed 14 December 2010). A revised version of the piece was subsequently published under the same title in *Interdisciplinary Science Reviews* 30, no. 2 (2005): 119–135.

Acknowledgments

William Aspray was the original editor of this volume, and his energetic commitment brought the book much closer to existence. He wrote its outline proposal, made an initial selection of work to include, and obtained reprint permissions from publishers. He and Joseph November also provided me with detailed comments on an early draft of its introduction. Jean Mahoney and Angela Creager commented on a late draft. My thinking on Mahoney's work has been guided by conversations with several other colleagues, including Gerard Alberts and Chigusa Kita. Edward Benoit, my research assistant, redrew all the diagrams from Mahoney's original published versions, and reworked citation styles to match the needs of Harvard University Press. Another assistant, Liliana Richard, carefully checked page proofs against the previously published text of each chapter and made many corrections. Michael Aronson at Harvard University Press propelled the book through various difficulties to rapid publication, while Melody Negron coordinated its production with an impressive eye for detail. The introduction was written during my time as a participant in the European Science Foundation's Software for Europe project.

"The History of Computing in the History of Technology" by Michael Sean Mahoney was originally published in *Annals of the History of Computing* 10 (April 1988), copyright © 1988 IEEE. Reprinted with permission.

"What Makes History?" by Michael Sean Mahoney was originally published in Thomas Bergin and Richard Gibson, eds., *History of Programming Languages*, vol. 2, pp. 831–832, copyright © 1996 ACM Press. Reproduced by permission of Pearson Education, Inc.

"Issues in the History of Computing" by Michael Sean Mahoney was originally published in Thomas Bergin and Richard Gibson, eds., *History of Programming Languages*, vol. 2, pp. 772–781, copyright © 1996 ACM Press. Reproduced by permission of Pearson Education, Inc.

"The Histories of Computing(s)" by Michael Sean Mahoney was originally published in *Interdisciplinary Science Reviews* 30, no. 2 (2005), pp. 119–135, copyright © Maney Publishing. www.maney.co.uk/journals/isr. Reprinted with permission.

"Software: The Self-Programming Machine" by Michael Sean Mahoney was originally published in A. Akera and F. Nebeker, eds., *Authoritative History of Modern Computing*, 2002. Reprinted with permission of Oxford University Press, Inc.

"The Roots of Software Engineering" by Michael Sean Mahoney was originally published in *CWI Quarterly* 3, no. 4 (1990), pp. 325–334.

"Finding a History for Software Engineering" by Michael Sean Mahoney was originally published in *Annals of the History of Computing* 26, no. 1 (January 2004), copyright © 2004 IEEE. Reprinted with permission.

"Boys' Toys and Women's Work: Feminism Engages Software" by Michael Sean Mahoney was originally published in Angela N. H. Creager, Elizabeth Lunbeck, and Londa Schiebinger, eds., *Feminism in Twentieth-Century Science, Technology, and Medicine*. Chicago: University of Chicago Press, 2001.

"Computing and Mathematics at Princeton in the 1950s" by Michael Sean Mahoney was originally published in *Les Cahiers de Science et Vie* 53 (October 1999).

"Computer Science: The Search for a Mathematical Theory" by Michael Sean Mahoney was originally published in John Krige and Dominique Pestre, eds., *Science in the 20th Century*, ch. 31. Amsterdam: Harwood Academic Publishers, 1997. Reprinted with permission.

"Computers and Mathematics: The Search for a Discipline of Computer Science" by Michael Sean Mahoney was originally published in J. Echevarria et al., eds., *The Space of Mathematics*, pp. 349–363. New York: De Gruyter, 1992. Reprinted with permission.

"The Structures of Computation and the Mathematical Structure of Nature" by Michael Sean Mahoney was originally published in the *Rutherford Journal* 3 (2008). Reprinted with permission.

"Software as Science—Science as Software" by Michael Sean Mahoney was originally published in Ulf Hashagen, Reinhard Keil-Slawik, and Arthur Norberg, eds., *History of Computing: Software Issues*. Berlin: Springer-Verlag, 2002. Reprinted with kind permission of Springer Science and Business Media.

"Éloge: Michael Sean Mahoney, 1939–2008" by Jed Z. Buchwald and D. Graham Burnett was originally published in *Isis* 100 (September 2009), pp. 623–626. Reprinted with permission from the University of Chicago Press, Jed Z. Buchwald, and D. Graham Burnett.

Index

Page numbers followed by *f* or *t* indicate figures or tables.

Abbott, Andrew, 5, 52–53
Adam, Alison, 223n11
Advanced Research Projects Agency (ARPA). *See* Defense Advanced Research Projects Agency (DARPA)
Agar, Jon, 64
Agendas: history of computer science, 130–131; history of computing, 52–54; history of software engineering, 92–93; research and training, 183–186, 188–190; theoretical computer science, 163–165, 164*f*, 167*f*, 185–188
Agresti, William W., 99–100
Aho, Alfred, 114, 169
Aitken, H. G. J., 205n3, 220n25
Algebra: Boolean, 132–133, 135, 136, 139, 148–149, 176, 187, 227n14; computer science as pure and applied, 14, 175–178, 176*f*, 234n34; modern versus classical, 158–159, 162
Algebraic machine theory, 141, 149, 227n4
"Algebraic Theory of Context-Free Languages, The" (Chomsky), 169
ALGOL, 31, 46–48, 58, 60, 81–82, 139–140, 150, 151–152, 167, 172, 174, 175, 211n25, 229n11
American system, 24–25, 28–29, 34. *See also* Mass production, used as paradigm
Annals of Mathematics Studies, 124, 127

Annals of the History of Computing. See IEEE Annals of the History of Computing
Application software, 78, 80–82, 215n4, 216n5
Applied Abstract Algebra (Lidl and Pilz), 158
Applied mathematics: agendas and theoretical computer science, 163–165, 164*f*; applied science and, 159–162, 230n6; history and, 162–163
Applied science: applied mathematics and, 159–162, 230n6; software engineering construed as, 93–96, 116, 185
Architecture, as model for history of software engineering, 104–105, 105*f*
Arden, Bruce, 144
Art of Computer Programming (Knuth), 130
Artifacts: applied science and applied mathematics, 159–162; computers and programs as, 183, 194; role in history of computing, 46–50, 68–70, 215n20
Artificial intelligence, 23, 28, 54, 93, 124, 174, 180, 189, 230n6; funding and, 30–31, 49, 189; humanities and, 72; LISP and, 127
Artificial life movement, 14, 69, 160
Aspray, William, 3, 49, 64, 241
Assembly lines. *See* Mass production, used as paradigm; Ford Model T, used as paradigm

Association for Computing Machinery (ACM), 53, 129; History of Programming Languages conferences, 79; SIGGRAPH, 54; SIGSOFT Impact Project, 13

Automata, Languages, and Machines (Eilenberg), 141, 169, 177–178

Automata, theory of, 12, 114, 134–138, 148–150; agendas and theoretical computer science, 163–165, 164*f*; formal languages and, 138–141, 166, 167*f*, 168–169; Princeton University and, 123, 124–126; von Neumann and, 133–134

Automata Studies, 166

Automatic programming, 84, 85, 103, 104, 188

Babbage, Charles, 5, 26, 30, 62, 148

Backus, John, 107, 139, 150

Bacon, Francis, 160

Bakker, Jaco de, 127, 188

Balzer, Robert, 103

Bartee, Thomas, 158–159, 176

Bashe, Charles, 29, 49

"Basis for a Mathematical Theory of Computation, A" (McCarthy), 93, 151, 170, 174

Bauer, F. L., 96, 98, 169, 188

Bell Labs, 9, 40, 49, 104, 110

Bemer, R. W., 84–85, 98–99

Berkeley, Edmund C., 54, 56

Between Human and Machine (Mindell), 63

Birkhoff, Garrett, 158–159, 162, 176–177, 229n1

Black-boxing, 135

BNF notation, 139, 150, 168

Boehm, Barry, 54, 77, 95–96, 219n16

Bolter, Jay, 71, 73

Boole, George, 5, 26, 148

Boolean algebra, 132–133, 135, 136, 139, 148–149, 176, 187, 227n14

Boulton, Matthew, 62

Bourbaki group, 137, 168, 178

Bowden, B. V., 88

"Boys' Toys and Women's Work: Feminism Engages Software" (Mahoney), 106–117; computers made masculine, 104–110, 222nn8,9; Haigh on, 10–11; literature case studies, 110–115; software engineering and, 115–117

Brandon, Dick H., 101, 220n30

Brooks, Frederick P., 66–67, 101, 116, 193, 237n19

Buchholz, Werner, 80

Büchi, J. Richard, 175

Burke, J. G., 210n19

Burks, Arthur, 59, 165

Burroughs, 55, 63, 200

Burstall, Rod, 172, 173–174, 175–176

Business. *See* Commerce and commercialization

C and C++, 79, 110

"Calculation—Thinking—Computational Thinking: Seventeenth-Century Perspectives on Computational Science" (Mahoney), 15

Campbell-Kelly, Martin, 61, 64

Capital (Marx), 213n1

"Capital Intensive Software Technology" (Wegner), 34

CASE (computer-assisted software engineering), 85, 103

Category theory, 127, 162, 169–170, 175–176, 187

Cellular automata, 14, 165, 178–182, 180, 181*t*

Chomsky, Noam, 13, 138–141, 150, 168–169, 231nn16,17

Church, Alonzo, 12, 126, 127, 132, 143, 151, 172

Circuit theory, 134–138, 149, 187

COBOL (Common Business Oriented Language), 31, 79, 81

Coding theory, 135, 136–138, 188

Cohen, I. Bernard, 205n3

Cohn, Paul M., 175, 176, 234n34

Commerce and commercialization: history of technology and creation of demand, 28–30; role in history of computing, 47–50; roots of software engineering, 88–89

Communications industry community, 62

Communities of computing, 59, 61–65, 62*f*

Comparable functions, in history of computer science, 131–134

Compiler, 46, 66, 81–82, 84–85, 89, 96, 102, 193, 238n20; theory of, 85, 114, 140, 150–151, 155, 172–176

"Computer revolution," declarations of, 56

Computer science, in tripartite nature of computing, 25–27

Computer Science and Engineering Research Study (COSERS), 191–192, 237n15

"Computer Science: The Search for a Mathematical Memory" (Mahoney), 128–146; comparable functions, 131–134; formal language theory, 138–141; formal semantics, 141–144; Haigh on, 13; limits of mathematical computer science, 144–146; theory of automata, 134–138

"Computers and Mathematics: The Search for a Discipline of Computer Science" (Mahoney), 147–157; Haigh on, 13, 147; McCarthy and, 151–153, 154; Scott and, 154–156; Strachey and, 147–148, 151, 153–154

"Computing and Mathematics at Princeton in the 1950s" (Mahoney), 121–127; artificial intelligence, 123, 124–127; game theory, 123, 124; Haigh on, 12; IAS computer, 121–123; lambda calculus, 126–127; non-linear computations, 123–124

Context-free grammars, 138–140, 150, 168–169, 178

Cook, Steven, 149

Cortada, James, 61, 63, 64, 78

Cowan, Ruth Schwarz, 25, 215n19

Cox, Brad, 97

Coy, Wolfgang, 238n22

CPL (Combined Programming Language), 149, 171–172, 174

Creager, Andrea, 14

Creating the Computer (Flamm), 48–49

Curry, Haskell, 170

Cusumano, Michael, 102, 103–104

Daniels, George S., 24, 28–29, 34, 209n5

Data processing community, 61

Datamation, 77

Datatron, 55, 200

Deductive approach, to semantics, 150

Defense Advanced Research Projects Agency (DARPA), 31, 49, 103

Denning, Peter, 230n6

Dennis, Jack B., 97

Denotational semantics, theory of, 127, 143

Descartes, René, 1, 197, 201

Diebold, John, 30, 79

Dijksta, Edsgar W., 105, 218n7

Documentation of practice, as issue in history of computing, 44–47

"Draft Report on the EDVAC" (von Neumann), 122

Drucker, Peter, 28–29, 100

Dubriel, Pierre, 137

Dynabook, 111

Eckert, J. Presper, 122, 216n2

EDSAC (Electronic Delay Storage Automatic Calculator), 70–71, 81

"Education of a Computer" (Hopper), 51

EDVAC (Electronic Discrete Variable Automatic Computer), 58f, 59, 61f

Eilenberg, Samuel, 141, 169, 177–178

Electrical engineering, in tripartite nature of computing, 25–27

Electronic accounting machinery (EAM), of IBM, 80–81

Elements of Programming Style (Kernighan and Plauger), 47

Éloge, to Mahoney, 197–202

ENIAC (Electronic Numerical Integrator and Computer), 5–6, 10, 26, 58f, 59, 60f, 122; women and, 106–107, 221n3

Entscheidungsproblem, 131–132

Erb, Ulrike, 113–115, 116–117

Estrin, Thelma, 110–111, 223n10

"Exploring the Excluded: A Feminist Approach to Opening New Perspectives in Computer Science" (Erb), 113

Feminism. *See* "Boys' Toys and Women's Work: Feminism Engages Software" (Mahoney)

Ferguson, Eugene S., 24, 25

Fermat, Pierre de, 199

"Finding a History for Software Engineering" (Mahoney), 56–57, 90–105; applied science and, 93–96; architectural model, 104–105, 105f; Haigh on, 8–9, 13; industrial engineering and, 98–104; mechanical engineering and, 96–97, 116; self-definition tasks and, 90–93, 217n3

Finerman, Aaron, 190

"Finite Automata and Their Decision Problems" (Rabin and Scott), 126, 133, 166

"Firsts," in history of computing, 4, 22, 27, 38, 43–45

Flamm, Kenneth, 48–49

Fontana, Walter, 178–179

Ford, Henry, 25, 45–46, 48, 49, 62, 68, 102–103

Ford Model T, used as paradigm, 8–9, 35, 36, 44, 48, 51–52, 54, 102–103. *See also* Ford, Henry; Mass production, used as paradigm

Formal language theory, 138–141, 142, 150, 180
Formal Languages and Their Relation to Automata (Hopcroft and Ullman), 140–141
Formal semantics, 82, 141–144, 169–174, 171*f*
Forsythe, George, 184, 189
FORTRAN, 31, 47–48, 61, 81
FUNARG problem, 228n10
Funding. *See* Government support

Game theory, Princeton University and, 123, 124
Gauss, Carl Friedrich, 198
Gender issues. *See* "Boys' Toys and Women's Work: Feminism Engages Software" (Mahoney)
General and Logical Theory of Automata (von Neumann), 166
Giant Brains, or Machines That Think (Berkeley), 56
Gillispie, Charles, 198
Ginsburg, Seymour, 139, 168
Go-Between, The (Hartley), 41
Gödel's dilemma, 126
Government Machine, The (Agar), 64
Government support, 48–49; agendas and, 188–189; for artificial intelligence, 30–31, 49, 189; for research and history of technology, 25, 30–31; for science at Princeton University, 121–122, 123
Graham, Ronald, 217n3
Graphical user interfaces (GUIs), 71, 79–80
Grusin, Richard, 71, 73

Hackers, 112–113
Haigh, Tom, 63–64
Hamming, Dick, 38, 46, 86, 87, 189
Handbook of Mathematical Psychology, 169
Handbook of Theoretical Computer Science, 194–195
Håpnes, Tove, 112–113
Hartley, Leslie, 41
Hartmanis, Juris, 149
Harvard Business Review, 29–30, 50
Hawking, Stephen, 160, 196
Herrmann, Cyril C., 30
Hidden Order: How Adaptation Builds Complexity (Holland), 182
Hilbert, David, 131–132, 162

Hindle, Brooke, 25
"Histories of Computing(s)" (Mahoney), 55–73; Haigh on, 5–7, 10; humanities applications and, 70–73; machine-centered view, 57–59, 58*f*, 60*f*, 61*f*; software design and operative representation, 65–68; software's accessibility as artifacts, 68–70; users and communities of computing, 59, 61–65, 62*f*
"History of Computing in the History of Technology" (Mahoney), 21–37; complex, tripartite nature of computing, 25–27; computing's present history, 21–23; Haigh on, 2–4, 205n2; history of computing as history of technology, 27–35; questions of history of technology, 23–25
History of Science Society, 3, 22, 43
Hoare, C. A. R., 94, 192–193
Hoare, Tony, 12, 194
Holland, John H., 182
Hopcroft, John, 114, 140–141, 169
Hopper, Grace, 9, 44, 51, 106–107, 221n2
Hounshell, David A., 24
Huffman, David A., 149
Hughes, Thomas P., 24
Humanities, applications of computing and, 70–73
Humphrey, Paul, 179
Humphrey, Watts S., 100
Huygens, Christiaan, 1, 14, 197

IBM, 29, 61, 63–64, 79, 80–81, 89
IBM Journal of Research and Development, 49
IBM's Early Computers (Bashe), 22, 29, 49
IEEE Annals of the History of Computing, 2, 10, 205n2
IEEE Software, 34, 51, 97
"Impact theory," of relation of technology and science, 57
Indiscrete Thoughts (Rota), 234n34
Industrial engineering: industrial engineering community, 62; software engineering construed as, 98–104, 116
Informatik, 195, 238n22
Information Technology as Business History (Cortada), 63
Ingrao, Bruna, 162–163
Institute for Advanced Science (IAS), 121–123

Interactive, real-time systems, government support for, 30

Into the Heart of the Mind (Rose), 30–31

Isis, 3

Israel, Giorgio, 162–163

"Issues in the History of Computing" (Mahoney), 42–54; agenda for history of computing, 52–54; context and role of commerce, 47–50; current practice and relation to history, 50–52; Haigh on, 4–5, 8–9; practice documentation needed, 44–47

Japan's Software Factories (Cusumano), 103–104

Jenkins, Reese, 25

Jones, Cliff, 207n15

Jones, Gregory W., 51, 97

Journalists, "social impact statements" and history of computing, 23

Karp, Richard, 149

Kay, Alan, 46–47, 70, 111

Kernighan, Brian, 47

Kidder, Tracy, 29, 33, 45, 69, 210n18

Kleene, Stephen C., 126, 132, 135–136, 150, 166, 168

Kling, Rob, 57

Knowledge, history and "what it was like not to know what they now know," 38–41

Knuth, Donald, 130, 184, 226n2

Kuhn, Thomas, 4–5, 16, 45, 86–87, 189, 199, 205nn4,5

Lambda calculus, 13–14, 132, 143, 154, 156, 169–170, 171f, 172–173, 179, 187, 188, 232n18; formal semantics and, 169–173, 175–176; mathematics and semantics of programming languages, 151–153; Princeton University and, 126–127

Landin, Peter J., 152, 153, 154, 172, 174, 175–176, 188, 232n18

Langton, Christopher, 69, 180

Lattice theory, 127, 139, 144, 158, 169, 176, 178; lattice-theoretical model of Scott, 156–157, 173, 187

Laws of software, 195–196

Layton, Edwin T., Jr., 24

Lefschetz, Solomon, 123, 124

Lidl, Rudolf, 158

Lindenmayer, Aristid, 178

Linux, 10, 112

LISP, 31, 81, 110–111, 127, 143, 152, 170, 223n11, 228n10

Logo, 110–111, 223n11

Machine-centered view, of history of computing, 57–59, 58f, 60f, 61f

Mack, Pamela, 106

MacKenzie, Donald, 207n15

MacLane, Saunders, 148

MacOS (Apple), 80

Magee, John F., 30

Management science community, 62

Marketplace. *See* Commerce and commercialization

Marx, Karl, 213n1

Mass production, used as paradigm, 35–36, 44, 96–97, 102–103

Mathematical Career of Pierre de Fermat, 1601–1655, The (Mahoney), 197, 199

"Mathematical Realm of Nature, The" (Mahoney), 200

"Mathematical Theory of Communication" (Shannon), 136

Mathematics: as applied science, 93–95, 159–162, 230n6; software engineering and science, 190–193, 237n15. *See also specific writings*

Mauchly, John, 122

McCarthy, John, 12, 218n10; applied science and, 93–94, 103, 228nn6,10; artificial intelligence, 124–125, 126–127; cellular automata, 180; formal semantics, 141–143; mathematics and semantics of programming languages, 151–153, 154, 170–172, 174–175, 229n11; theoretical computer science, 188, 189

McCarty, Willard, 72–73

McCulloch, Warren, 134, 135, 150, 166

McDougall, Walter, 25

McIlroy, M. D., 34, 35, 52, 96–97, 111, 219n18

McKenzie, Donald, 9

Mealy, G. H., 149

Mechanical engineering, software engineering construed as, 96–97

Mehrtens, Herbert, 162

Meyer, Stephen, 25

Microelectronics, government support for, 30

Military command and control community, 63

Miller, George, 139, 168, 231n16
Mindell, David, 63
Minsky, Marvin, 12, 124, 130, 184, 189, 226n2
Model T Ford. *See* Ford Model T
Modern Applied Algebra (Birkhoff and Bartee), 158–159
Moore, E. F., 149, 166
Morgenstern, Oskar, 123
Morison, Elting E., 210n16
Moses, Joel, 224n23
Mythical Man-Month, The (Brooks), 66, 101

Nash, John, 123
National Cash Register (NCR), 63
National Initiative to Network the Cultural Heritage (NINCH), 72
NATO Science Committee, 32, 87, 92
NATO Software Engineering Conference (1968), 9, 34, 52, 84, 92, 116, 206n9
NATO Software Engineering Conference (1969), 93, 95, 116
Nelson, Daniel, 25
Nelson, Ted, 50
Nerode, Anil, 154, 229n11
Networks of Power (Hughes), 24
New Kind of Science, A (Wolfram), 160
Newell, Allan, 129–130, 184, 226n2
Newton, Isaac, 1, 17, 93, 95, 140, 142, 171, 180, 198
Noble, David. F., 25
Non-linear computation, 122–124, 128, 163, 179
Norberg, Arthur, 49
Numerical analysis, 14, 53, 128, 149, 163
Nygaard, Kristen, 48

Object-oriented programming, 52, 97, 110–111
Oettinger, Anthony, 169
Office of Naval Research (ONR), 123
"On a Logical and General Theory of Automata" (von Neumann), 133–134
"On an Application of Semi-Group Methods to Some Problems in Coding" (Schützenberger), 166
"On Mass-Produced Software Components" (McIlroy), 52
On the Economy of Machinery and Manufactures (Babbage), 62
O'Neill, Judy, 49

Operating systems. *See* Systems software
Operational approach, to semantics, 150
Operations research: history of technology and, 29–30; operations research community, 62
Operative representation, software design and, 65–68
OS/360 (IBM), 83
Osterweil, Leon J., 100
"Outline of a Mathematical Theory of Computation" (Scott), 143–144

Papert, Seymour, 110
Penrose, Roger, 172
Perlis, Alan, 47–48, 129–130
Perrin, Dominique, 169
Petroski, Henry, 66
Piaget, Jean, 38–39
Pierce, John, 184
Pilz, Gunter, 158
"Pipes," 113, 219n18
Pitts, Walter, 134, 135, 150, 166
Plauger, P. J., 47
"Position Paper for Panel Discussion [on] the Economics of Program Production" (Bemer), 98–99
Post, Emil, 139, 168
Price, Derek J. de Solla, 47, 68, 87
Princeton Logistics Project, 123
Princeton University, 49, 71–72. *See also* "Computing and Mathematics at Princeton in the 1950s" (Mahoney)
Problem-oriented languages. *See* Application software
Programming languages: automata theory and, 166, 167f, 168–169; history of technology and, 31, 211n25. *See also* Semantics
Psychology of Computer Programming, The (Weinberg), 47

Rabin, Michael, 126, 133, 135–136, 149, 150, 166, 169
RAND Corporation, 124
Randell, Brian, 216n2
Remediation (Bolter and Grusin), 73
Research and training, science agendas and, 188–190
Research funding. *See* Government support
Rheingold, Howard, 31

Rice, H. G., 139, 168
Ritchie, Dennis, 112
Rochester, Nathaniel, 124, 152, 170, 189
"Roots of Software Engineering, The"
  (Mahoney): commercialization, growth
  of programming, emergence of software,
  88–89; Haigh on, 8, 86; software-
  engineering precedents, 86–87
Rose, Frank, 30–31
Rosenberg, Nathan, 24, 27, 35, 97
Rosenbloom, Paul C., 168, 232n18
Rosin, Bob, 51
Ross, D. T., 216n2
Rota, Gian-Carlo, 234n34
Russell's paradox, 126, 143
Rutishauser, Heinz, 188, 215n4

Samelson, Klaus, 169, 188
Sammet, Jean, 11, 107
Sarnoff Laboratory, of RCA, 122
Scacchi, Walt, 57
Schützenberger, Marcel P., 13, 163, 182,
  188; automata theory and, 137–138,
  166, 168–169; context-free grammars
  and, 139, 140
Schwarzschild, Martin, 122
Science. See Applied science; specific writings
Sciences of the Artificial, The (Simon), 209n9
"Scientific management," 99–101, 220n25
"Scientific revolution," 12, 14, 16–17, 56,
  68, 197–198
Scott, Dana, 12, 126, 127, 188; automata
  theory and, 133, 135–136, 166, 168;
  lattice-theoretical model of, 156–157,
  173, 187; mathematics and semantics of
  programming languages, 143–144, 150,
  154–156, 173, 174, 232n18, 233n26
Self-programming computer. See
  "Software: The Self-Programming
  Machine" (Mahoney)
Semantics: denotational, 127, 143; formal,
  82, 141–144, 169–174, 171f; lambda
  calculus and, 127, 169–173, 175–176;
  mathematics and semantics of programming languages, 147–157, 170–172,
  174–175, 229n11; operational approach
  to, 150
Shannon, Claude, 124, 132–133, 135, 136,
  148–149, 189, 196
Shapiro, Ehud, 230n6
Sharp, I. P., 104
Shaw, Mary, 51, 217n3, 224n22

Shubik, Martin, 124
Simon, Herbert, 30, 79, 129–130, 183,
  184, 194, 209n9, 226n2
Simula, 46–47, 70, 110, 111
Smalltalk, 111
Smith, Merritt Roe, 24
"Social impact statements," 23
Society for the History of Technology, 3, 4,
  22, 43
Software: laws of, 195–196; use of term,
  193–194, 238n20
"Software and Its Impact: A Quantitative
  Assessment" report, 77
"Software as Science—Science as Software" (Mahoney), 183–196; agendas
  and institutions in computer science,
  183–190; criticisms of, 159; Haigh on,
  15–17, 183, 190–191, 195; laws of
  software, 195–196; mathematics and
  software engineering, 190–193, 237n15;
  software as science, 193–195
"Software crisis," 9, 31–32, 53, 66, 77, 87,
  206nn6,9
Software engineering: agenda for history of
  computing and, 52–54; government
  support for, 30; history of technology
  and, 27–28, 31–36; inaccessibility and
  nondocumentation of software, 42; origin
  of term, 32, 87, 216n2; software design
  and operative representation, 65–68; in
  tripartite nature of computing, 25–27
Software Engineering (Jones), 51, 97
Software Engineering Notes, 54
"Software factory," 80, 84–85
"Software: The Self-Programming
  Machine" (Mahoney), 77–85; Haigh on,
  7–8; programming tools and environments, 78, 80–82, 215n4, 216n5;
  "software factory" and programmers,
  80, 84–85; systems software, 82–84
Sørensen, Knut, 112–113
Soul of a New Machine (Kidder), 29, 33,
  45, 69, 210n18
Stack, notion of, 140, 169
Stearns, Richard E., 149
Strachey, Christopher, 12, 13, 95, 127, 188;
  mathematics and semantics of program-
  ming languages, 147–148, 151,
  153–154, 172–173, 191
Stroustrup, Bjarne, 51
Structure of Scientific Revolutions, The
  (Kuhn), 16, 45, 86–87

"Structures of Computation and the Mathematical Structure of Nature, The" (Mahoney), 158–182; applied mathematics and its history, 162–163; applied science and applied mathematics, 159–162, 230n6; automata and formal languages, 166, 167f, 168–169; cellular automata and, 178–180, 181t, 182; computer science as pure and applied algebra, 175–178, 176f; formal semantics, 169–175; Haigh on, 14–15; modern versus classical algebra, 158–159, 162; von Neumann and theoretical computer science, 163–165, 164f

Structures of Computation: Mathematics and Theoretical Computer Science, 1950–1970, The (Mahoney), 11

Suchman, Lucy, 115, 117

Switching circuits. *See* Circuit theory

Symbolic assemblers, 81

Symposium on Information Theory (MIT), 166, 168, 231n16

System of Professions, The (Abbott), 52–53

Systems Development Corporation, 102

Systems software, 7, 78–80, 82–84

Tarski, Alfred, 169

Taylor, Frederick W., 9, 30, 62, 99–101, 220n25

Theoretical computer science, 32–33, 185–188, 191

Theory of Games and Economic Behavior (Morgenstern and von Neumann), 123

"Theory of Neural-Analog Reinforcement Systems" (Minsky), 124

Thompson, Ken, 112

"Three Models of Language" (Chomsky), 138–139

Tomayko, James, 66

Tools for Thought (Rheingold), 31

"Toward a Formal Semantics" (Strachey), 154

"Toward a Mathematical Semantics for Computer Languages" (Strachey and Scott), 156

"Towards a Mathematical Science of Computation" (McCarthy), 154, 170, 174

Training, science agendas and, 188–190

Tukey, John, 194

Turing, Alan, 11, 13, 63, 78–79, 106, 126, 131–132, 133, 148, 196

Turing Award, 129, 226n2, 235n3

Turing machine, 28, 59, 78–79, 87–88, 109, 133, 148–149, 150, 196, 201

Turkle, Sherry, 110, 112, 113, 223n14

Ulam, Stanislaw, 125

Ullman, Jeffrey, 114, 140–141, 169

Undocumented practice, as issue in history of computing, 44–47

"Une théorie algébrique du codage" (Schützenberger), 137–138

Universal Algebra (Cohn), 175, 176, 234n34

University Education in Computing Science (Finerman, ed.), 190

UNIX, 9–10, 79, 104, 112–113, 114

Users, communities of computing and, 59, 61–65, 62f

Vogel, Kurt, 198

Von Neumann, John, 11, 12, 26, 63, 79; agendas and theoretical computer science, 163–165, 164f; game theory, 123, 124, 162–163; IAS computer and, 121–123; non-linear computations, 123–124; theory of automata, 133–134, 166, 178

Wajcman, Judy, 109

Wallace, Anthony F. C., 25

Watt, James, 62

"We Would Know What They Thought When They Did It" (Hamming), 38, 86, 87

Wegner, Peter, 34, 51, 97, 184, 230n6

Weinberg, Gerald M., 33, 47

Weizenbaum, Joseph, 112

West, Tom, 33, 45, 69

What Can Be Automated?, 144

"What Makes History?" (Mahoney), 38–41

"What Makes the History of Computing Hard and Why Does It Matter?" (Mahoney), 10

Wigner, Eugene, 122, 160

Wilkes, Maurice, 80

Wilks, Yorick, 65

Williams, S. B., 238n20

Windows NT (Microsoft), 83

Winner, Langdon, 68, 215n19

Wolfram, Stephen, 160, 180

"Women's Studies and Computer Science: Their Intersection" (Estrin), 110

Yates, JoAnne, 64

Harvard University Press is a member of Green Press Initiative
(greenpressinitiative.org), a nonprofit organization working to
help publishers and printers increase their use of recycled paper
and decrease their use of fiber derived from endangered forests.
This book was printed on recycled paper containing 30%
post-consumer waste and processed chlorine free.